Calculating

Trade Union Industrial Studies

This series makes two new types of provision in the area of industrial relations: first it is specifically directed to the needs of active trade unionists who want to equip themselves to be more effective, and second, the books are linked together in a series related to the requirements of existing training and education courses.

The books have been designed by a Curriculum Development Group drawn from the Society of Industrial Tutors: Michael Barratt Brown, Ed Coker, Bob Houlton, Ossie O'Brien and Geoffrey Stuttard, together with Charles Clark and Francis Bennett of the Hutchinson Publishing Group. The Curriculum Development Group has prepared the guide lines for the texts and edited them so that they form a complete set of teaching material for tutors and students primarily for use on trade union courses.

The texts are issued in sets of four, together with an accompanying resource book which provides additional background material for tutors and students.

Trade Union Industrial Studies

This series is published in three sets, each consisting of four student texts and an accompanying resource book. This book includes additional teaching material for tutors and students, a recommended list of books and a further exploration of the subjects by the authors of the texts.

The Activist's Handbook   *Bob Houlton*
Statistics for Bargainers   *Karl Hedderwick*
Calculating   *Joyce & Bill Hutton*
The Organized Worker   *Tony Topham*
Industrial Studies 1: the key skills
Eds. Ed Coker & Geoffrey Stuttard

Industrial Action   *Ernie Johnston*
Pay at Work   *Bill Conboy*
Management Techniques   *Jim Powell*
The Job of the Manager   *John Bailey*
Industrial Studies 2: the bargaining context
Eds. Ed Coker & Geoffrey Stuttard

The Union Rule Book   *Richard Fletcher*
Opening the Books   *Michael Barratt Brown*
Trade Unions and Government   *Ossie O'Brien*
Workers' Rights   *Paul O'Higgins*
Industrial Studies 3: understanding industrial society
Eds. Ed Coker & Geoffrey Stuttard

Joyce and
Bill Hutton

Calculating

Hutchinson
in association with the
Society of Industrial Tutors

Hutchinson & Co (Publishers) Ltd
3 Fitzroy Square, London W1

London Melbourne Sydney Auckland
Wellington Johannesburg Cape Town
and agencies throughout the world

First published 1975
© Joyce and Bill Hutton 1975

Set in Press Roman by E.W.C. Wilkins Ltd
London and Northampton
Printed in Great Britain by litho by The Anchor Press Ltd
and bound by Wm Brendon & Son Ltd
both of Tiptree, Essex

ISBN 0 09 122880 8

Contents

Introduction	9	Percentages	39
Guidelines	9		

Chapter 1

		Chapter 4	
Powers of 2	11	Powers of 2	42
Using powers of 2	12	Binary number	43
Fractions	12	Fractions	44
Ratios	14	Decimal fractions	45
Decimal fractions	15	Graphs	46
Percentages	17	Significant figures	48
A simple slide rule	19	The commercial slide rule	48

Chapter 2

		Chapter 5	
Computer numbers	21	The theory of indices	52
Powers of 2	23	Ratio and proportion	54
Division	24	Estimations	56
Fractions	24	Formulae	57
Changing denominators to powers of 10	25	Square roots	57
		The slide rule	59
Ratios	28	Interpreting graphs	60
Round figures	28		
The slide rule	29	**Chapter 6**	

Chapter 3

		Powers of 2	62
Negative numbers	31	Fractions: division	63
Powers	32	Graphs	64
Decimal fractions and degrees of accuracy	33	The slide rule	65
		Calculators	67
Recurring decimals	35		
Ratios as decimals	36	**Chapter 7**	
The decimal point	37	More powers	68
Computer numbers	38	Decimals: four rules	69
		The slide rule	72

Chapter 8

Logarithms	75
Ratio and proportion	76
Percentages: up and down	77
Simple interest	78
The slide rule	81

Chapter 9

Negative numbers	84
The slide rule	86
Ratio and proportion	87
Graphs	89
Postscript	91
Answers	93

Acknowledgements

We are indebted to many friends who gave us information about the intricacies of their wage packets. This background knowledge has been a great help to us. So too has been the help of those who agreed to be guinea-pigs and tried out the first draft of the book. Their comments, criticisms and corrections have been gratefully received and used to produce this work.

Introduction

If you feel that your maths is not good enough to cope with problems about wage rates, income tax, percentages and so on, this book is for you.

Here are tools for taking the hard grind out of arithmetic. No tool is more reliable than the person who uses it. Maths needs more than skill with numbers. The most important part is logic. You start with a practical problem. When you pick up your slide rule you must decide what to do with it. You must then decide whether the figures you read off are your wages, the pig meat production for 1973 or the National Debt. To begin this book you need to know your two times table. From then on you will learn the clever tricks invented to make calculating easy.

Guidelines

1. Don't try to swallow this book whole. It is very indigestible. Chew over each sentence. If you don't understand something talk it over with a friend.

2. Do *all* the drawings and exercises. There are answers at the back, but don't cheat yourself. If you find that some of your answers are wrong put them right before going on.

3. Here and there you will find some extra spacing, as shown here:

This means STOP! Think out for yourself what you have just read. Only then should you read the explanation that follows and see if you agree.

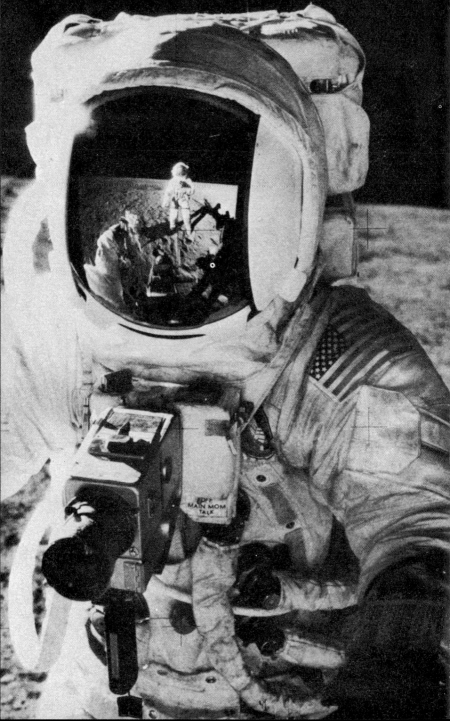

Chapter 1

This chapter begins with a very simple mathematical table which made possible the later development of logs, slide rules, computers and numerically controlled production systems. Without computers man's landing on the moon would have been impossible.

Powers of 2

Take a piece of paper. Fold it to make two equal parts. Fold again. Now you have four thicknesses of paper. Fold again and you will have...?

Our minds jump to 'six' because 'two, four, six, eight' is a familiar jingle.

Count the thicknesses of paper and you will find that there are not six but eight.

Remember that we are not adding two each time we fold, but doubling − multiplying by two.

These numbers form part of the set of numbers which make up the powers of 2

$$1, 2, 4, 8 \ldots$$

Notice that we start from 1, not 0. If you multiply 0 by 1 or 0 by 2 you will see why.

Now consider this statement. If you had a sheet of paper $\frac{1}{10}$ mm thick and could double it 20 times, the wad would then be over 100 metres thick. Do you believe this? Do you think it is absurd?

Make a table like this:

Thicknesses	*Number of times doubled*
1	0
2	1
4	2

Continue until you reach 20 doubles. You will now have a clue to the problem of the impossibly thick wad of paper. Keep the table you have made.

12 Calculating

Using powers of 2

You can use your table of powers of 2 to multiply.
  Here is an example: 16 × 32.
This could be written: 1 × 2 × 2 × 2 × 2 × 2 × 1 × 2 × 2 × 2 × 2 × 2;
which is the same as: 2 × 2 × 2 × 2   × 2 × 2 × 2 × 2 × 2,
or nine folds.

From your table you will find that this gives 512 thicknesses of paper. You may find it easier to write your working this way.

Thicknesses	Times folded
16	4
32	5
512	9

  Answer: 16 × 32 = 512

Exercise 1.1

Find:
(a) 128 × 64
(b) 2048 × 512
(c) 512 × 32
(d) 8 × 4
(e) 4096 × 128

Fractions

'Fraction' and 'fracture' come from the same Latin word meaning 'broken'. A fraction is a broken part of a whole one. Before looking at our own system it will be useful to examine the method used by an earlier civilization. The ancient Egyptians had a system with signs for a half, a third, a quarter, and so on, but they did not have our arrangement of two numbers, one above and one below a line. Quantities such as $\frac{9}{16}$ had to be written as 'a half and a sixteenth'. The only exception was $\frac{2}{3}$ for which they had a special sign. Their signs for a quarter were ⌒ and an eighth ⌒. Make up signs for other fractions.

> **Exercise 1.2**
> Show how the Egyptians might have written these fractions as an addition of two or more signs:
> (a) $\frac{3}{4}$ (d) $\frac{5}{8}$
> (b) $\frac{11}{16}$ (e) $\frac{15}{16}$
> (c) $\frac{7}{8}$

The Egyptians could no more add a half and an eighth than they could add an elephant and a water melon. They were different things. The system we use allows us to cut up one piece into smaller pieces. When all the pieces are of the same size we can add them. For example

$$\frac{1}{2} + \frac{1}{16} = \frac{8}{16} + \frac{1}{16} \quad \text{or} \quad \frac{8+1}{16}$$

$$= \frac{9}{16}$$

Similarly, we can subtract

$$\frac{1}{2} - \frac{1}{16} = \frac{8}{16} - \frac{1}{16} \quad \text{or} \quad \frac{8-1}{16}$$

$$= \frac{7}{16}$$

So long as we have the same-sized pieces we can add and subtract.

The name of the piece is the same in the answer as it was while we were adding or subtracting. 2 pence added to 7 pence is 9 pence. 8 plums take away 1 plum leaves 7 plums and 8 sixteenths take away 1 sixteenth leaves 7 sixteenths.

The bottom figure of a fraction is called the *denominator* because it names the pieces.

> **Exercise 1.3**
> Find:
> (a) $\frac{3}{4} + \frac{1}{8}$
> (b) $\frac{11}{16} - \frac{1}{4}$
> (c) $\frac{3}{32} + \frac{7}{16}$
> (d) $\frac{5}{32} + \frac{1}{2} - \frac{3}{16}$
> (e) $\frac{1}{4} + \frac{1}{2} - \frac{1}{32}$

14 Calculating

Draw a rectangle — say 16 cm × 8 cm. Divide it into two equal parts. Colour one half. Divide the other half giving two quarters of the original rectangle. Colour one of these. Repeat the process alternately using light and dark colours, as often as possible (see Figure 1.1).

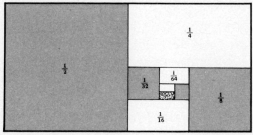

Figure 1.1

You will see from Figure 1.1 that however many fractions you add in this series, they will never add up to a total of more than 1. If we now put together the series of powers of 2 and this series of fractions, we have

$$\ldots 32, 16, 8, 4, 2, 1, \tfrac{1}{2}, \tfrac{1}{4}, \tfrac{1}{8}, \tfrac{1}{16} \ldots$$

We can extend this to the left indefinitely, getting larger and larger numbers. We can also continue indefinitely to the right getting smaller fractions and yet never reach 0 (zero). Each item is half of the number to the left of it throughout the whole series. Notice that the number 1 must be included to make this rule work.

Ratios

Fractions can be used to explain a relationship between two quantities. Take gear wheels for example.
In Figure 1.2 the relationship between the wheels is 9 inches to 3 inches, which we can write as a fraction

$$\frac{9 \text{ inches}}{3 \text{ inches}}$$

This means 9 inches divided by 3 inches.

The division sign ÷ is a fraction where the dots represent figures.

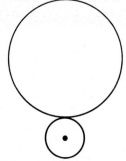

Figure 1.2

We could ask: 'How many times can a 3-inch length of wood be cut from a length of 9 inches?' The answer is a number, not a length – in this case, three. As long as the names of the units are the same at the top and bottom of the fraction the units can be cancelled. Therefore, in our wheel example we now have $\frac{9}{3}$. Both top and bottom can be divided by 3, giving $\frac{3}{1}$. This is *cancelling* and is used to simplify fractions. How does it work?

Take the same fraction, $\frac{9}{3}$, again. This could be written

$$\frac{3 \times 3}{1 \times 3} \quad \text{or} \quad \frac{3 \times 3}{1 \times 3}$$

Since 3 thirds is the same as a whole 1, the fraction can be written as $\frac{3}{1} \times 1$ which when multiplied gives $\frac{3}{1}$. This is the same as 3 whole ones. We call this a *ratio* of 3 to 1.

Decimal fractions

The earliest known use of fractions in Egypt is contained in the Rhind Papyrus, whose date is about 2000 BC. It was, as we have seen, a difficult process and was probably worked by a system of trial and error. The Babylonians used a system based on 60. If you turn the fractions $\frac{1}{2}$, $\frac{1}{3}$, $\frac{1}{4}$, etc., into sixtieths you will see that many of them work very simply.

Our system of common ('vulgar') fractions differs from that used by the Egyptians because we are not restricted to one of each size of piece. We can have any figure at the top of the fraction. **This top figure is called the numerator and shows how many pieces there are.**

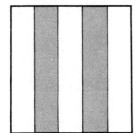

There are two ways of illustrating a fraction such as $\frac{2}{5}$. In Figure 1.3 a square has been divided into five pieces and two are shaded. This is two out of five, or $\frac{2}{5}$.

Figure 1.3

In Figure 1.4 there are two squares, and a fifth of each has been shaded, thus showing a fifth of two whole ones. The result is still $\frac{2}{5}$.

16 Calculating

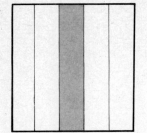 

Figure 1.4

It was more than three thousand years later, that is, about AD 1000, before a new system of fractions was discovered which made all fractions as easy to add and subtract as ordinary numbers.

The Egyptian idea of using two or more fractions — for instance, writing $\frac{1}{2} + \frac{1}{8}$ for $\frac{5}{8}$ — was adopted, but the denominators were limited to those based on 10 — that is tenths, hundredths, thousandths, etc. The numerator could be any number from 0 to 9, for instance $\frac{2}{10} + \frac{5}{100}$. The advantage of this system was that it fitted the ordinary number series based on 10. For example, $307 + \frac{2}{10} + \frac{5}{100}$ can be written

100	10	1	$\frac{1}{10}$	$\frac{1}{100}$
3	0	7	2	5

Of course, this can be confusing once we leave out the headings — unless we show where the whole numbers end and the fractions begin. Early in the seventeenth century, a famous mathematician, Napier, used this notation: 307 2' 5". In the same year his friend Briggs chose a simpler method: 307·25. This is the system we use in Britain today. (The Americans use 307.25 and the French 307,25.)

It is now an easy matter to change 41 tenths of an inch to 4.1 inches; 412 pence to £4·12 and 4125 millimetres to 4·125 metres.

Exercise 1.4

	Convert to inches:		Convert to £'s:		Convert to metres:
	Tenths of inches		*Pence*		*Millimetres*
(a)	73	(f)	543	(k)	6894
(b)	128	(g)	206	(l)	3204
(c)	4	(h)	59	(m)	7
(d)	326	(i)	1	(n)	532
(e)	155	(j)	875	(o)	68

Any empty spaces must be filled with a 0 if this helps to show the position of the number on the number line. We do this when we write 3p as £0·03, the form used in cash tills and calculators.

In Figure 1.5 two tenths and five hundredths have been shaded. In decimals that is 0·25. Draw diagrams using graph paper to show these fractions: 0·7, 0·4, 0·01, 0·21, 0·38.

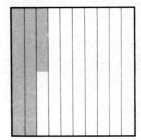

Figure 1.5

Percentages

In Latin *'per'* means 'for' (each) and *'cent'* is short for *'centum'* which means one hundred. In a factory of a hundred workers, three are doing maintenance. This is three out of a hundred, which is 3 per cent (3%). (See Figure 1.6.)

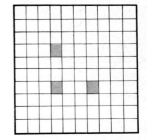

Figure 1.6

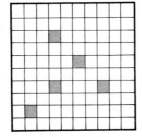

Figure 1.7

Figure 1.7 shows two workshops, each employing a hundred men, five of whom are foremen. Because there are five foremen in each hundred, this represents 5 per cent (5%).

Figure 1.8 shows four workshops each employing a hundred people. The black squares show the number of women. To find the number of women as a percentage we must find how many black squares would be in each hundred if they were equally spread — that

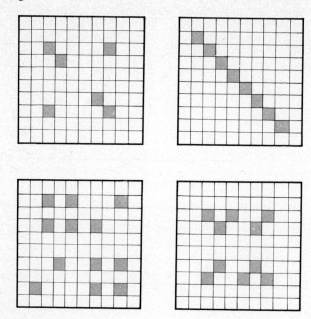

Figure 1.8

is we must find the *average*. There are thirty-six black squares altogether, which is nine black squares in each hundred. Therefore, in Figure 1.8, 9% are women.

A percentage is the average number in each hundred.

In Figure 1.9 we have four black squares out of fifty. A percentage always tell us how many out of each hundred. In this example it would not be difficult to guess how many squares will be black if there are a hundred squares and black ones appear as frequently as they do in Figure 1.9. There will be eight, so four out of fifty is 8%.

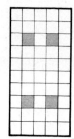

Figure 1.9

Exercise 1.5

Find the percentage:

(a) 5 out of 20
(b) 1 out of 5
(c) 17 out of 50
(d) 8 out of 10
(e) 13 out of 20
(f) 60 out of 200
(g) 80 out of 400
(h) 15 out of 300
(i) 90 out of 900
(j) 100 out of 1000

Common and decimal fractions, ratios and percentages are different ways of looking at the same thing: the relationship between two numbers.

A simple slide rule

Take a piece of paper. Cut it in half lengthways. Fold each strip lengthways so as to make a firm, straight edge. Mark off a number of equal spaces along the edge of one piece. Then mark off exactly similar spaces along the edge of the other piece. Number these graduations with the powers of 2 from left to right. Figure 1.10 shows how it should look.

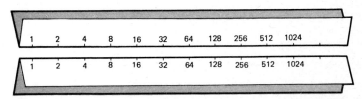

Figure 1.10

Suppose you want to multiply 64 by 8. Slide the top strip along until the 1 is over the 64 on the bottom strip. Look for the 8 on the top strip and the answer will be directly under it on the bottom strip. Figure 1.11 shows that $64 \times 8 = 512$.

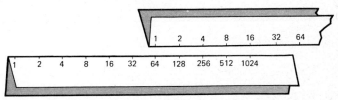

Figure 1.11

20 Calculating

You have added the three spaces between 1 and 8 on the top slide to the six spaces between 1 and 64 on the bottom slide. The result is nine spaces. If you look back to the section where we were considering the number of times a piece of paper was folded, you will see that the spaces on the slide rule correspond to the number of times the paper was folded.

A standard slide rule is more complicated than your strips of paper.

You will be able to find the numbers 1, 2, 4 and 8, and one scale will also have marks to show the position of 16, 32 and 64. If you can find them you can try out some simple multiplications as you did with your paper slide rule.

Chapter 2

Computers make calculations about a hundred million times faster than pencil and paper. Although the output will be in the kind of figures which we can readily understand, computers use powers of 2.

Computer numbers

We find powers of 2 in our old system of weighing in pounds and ounces. If we use a pair of scales we can weigh any number of ounces less than 1 lb without requiring weights other than 1 oz, 2 oz, 4 oz and 8 oz. For larger quantities we would need 1 lb and 2 lb weights and so on, that is, 16 oz, 32 oz, etc.

We can show how we would weigh any number of ounces less than 1 lb as follows:

		Weights		
8 oz	*4 oz*	*2 oz*	*1 oz*	*Total*
1	0	0	1	9 oz
0	1	1	0	6 oz
0	1	0	1	5 oz

The weights we use are marked by 1 and those we do not use by 0. Computers and production control systems use numbers of this kind but do not use the same headings. You may come across some numbers written like this

Base 2
1 011
1 101
0 111

22 Calculating

In these numbers the value of each figure 1 is decided by the *place* it occupies in the series of powers of 2. Can you write down the quantities in words instead of figures?

Exercise 2.1

Make a table showing the following in base 2 numbers:

(a) 3 oz
(b) 10 oz
(c) 15 oz
(d) 1 oz
(e) 12 oz

The heading Base 2 above is a warning that these are not ordinary numbers. They require headings similar to those in the table of ounces.

64	*32*	*16*	*8*	*4*	*2*	*1*
0	0	0	1	0	1	1

The number 1 011 (base 2) is $8 + 2 + 1$, that is, 11.

Now look at our ordinary number system. If we make a table of powers of 10 in the same way as we did with powers of 2, it will begin as follows

$$10^0 = 1$$
$$10^1 = 1 \times 10 = 10$$
$$10^2 = 1 \times 10 \times 10 = 100$$
$$10^3 = 1 \times 10 \times 10 \times 10 = 1000$$

These numbers can be arranged in a line as we did with ounces

1000	*100*	*10*	*1*
1	0	0	1

This is 1001, that is, one thousand and one. Because our ordinary number system is based on powers of 10 we have no difficulty in reading this off as 1001. We are accustomed to this so that we understand immediately how big a quantity is intended.

This is called the *decimal notation*. 'Decimal' means not only decimal fractions (numbers with a decimal point in them) but any number based on powers of 10.

Numbers based on the powers of 2 are called *binary numbers*. 'Bi' means 'two' as in 'bicycle'. The reason that the binary system is used in computers is that an electric switch can only be 'on' or 'off'. The 1 in binary number can be represented by a switch 'on' and the 0 by a switch 'off'. It is the only system which can represent any number with two figures, 1 and 0.

If you go back to paper folding you will remember that the twentieth power of 2 was equal to more than 1 000 000. How much, roughly, would be shown if all 20 switches were 'on'?

> **Exercise 2.2**
>
> Write these numbers according to the binary system. If it will help, put these headings as you did for weights but leave out the oz.
>
> (a) 11
> (b) 13
> (c) 4
> (d) 14
> (e) 7

Powers of 2

We saw how by folding a piece of paper we could make a series of numbers which we called the 'powers of 2'. When it became impossible to fold the paper any further we continued the series by multiplying. We were then working with *abstract numbers*.

Mathematicians show the number of times 1 has been doubled by writing 2^1, 2^2, 2^3, etc. These small figures, placed to the top right of the figure which is the base of our multiplication, are called *indices* and show which power each number is. For instance 512 is 2^9, that is 2 to the power of 9, or 2 to the ninth power. This notation makes it possible to write multiplications as follows

$$\begin{array}{rl} 512 & 2^9 \\ \times \ 128 & 2^7 \\ \hline 65\,536 & 2^{16} \end{array}$$

Answer: $512 \times 128 = 65\,536$

We add 9 and 7. The answer is then found from the table as before.

Division

Can you guess what the answer would be if we divided 512 by 128? If so, you will be able to find a rule for dividing using indices.

$$\begin{array}{cc} & 512 & 2^9 \\ \text{divided by} & \underline{128} & \underline{2^7} \\ & \rule{1cm}{0.4pt} & \rule{0.5cm}{0.4pt} \end{array}$$

Write down as simply as possible the rules you have found for multiplying and dividing using indices.

Exercise 2.3

Use your table of powers of 2 to work these examples and see if you get the answer you would expect:

(a) $32 \div 8$
(b) $1024 \div 512$
(c) $8192 \div 1024$
(d) $131\,072 \div 128$
(e) $262\,144 \div 64$

Fractions

Here are more fractions to add and subtract. These are not based on powers of 2.

$$\tfrac{1}{6} + \tfrac{3}{4}$$

Before we add, we must make the pieces the same size. Imagine that these are fractions of 1 ft. Then it would be simple to write these quantities as inches. $\tfrac{1}{6}$ ft is 2 inches; $\tfrac{3}{4}$ ft is 9 inches. Together they make 11 inches.

We could also say that an inch is 1 twelfth of 1 ft. We would then add 2 twelfths and 9 twelfths, which gives the answer: 11 twelfths. Using figures instead of words the whole calculation could be written

$$\begin{aligned} & \tfrac{1}{6} + \tfrac{3}{4} \\ =\ & \tfrac{2}{12} + \tfrac{9}{12} \\ =\ & \tfrac{11}{12} \end{aligned}$$

> **Exercise 2.4**
> Find:
> (a) $\frac{2}{3} + \frac{11}{12}$
> (b) $\frac{5}{12} - \frac{1}{3}$
> (c) $\frac{1}{2} - \frac{1}{12}$
> (d) $\frac{7}{12} + \frac{1}{2}$
> (e) $\frac{5}{6} + \frac{3}{4}$
> (f) $\frac{2}{3} + \frac{1}{4} + \frac{7}{12} + \frac{1}{2}$
>
> Some examples will add up to more than 12 twelfths and will make more than a whole 1. For instance, $\frac{17}{12}$ should be written as $1\frac{5}{12}$ (17 inches is 1 ft 5 in).

It is necessary when changing denominators to find a number into which all the denominators in the figures in front of you will divide exactly. The smaller numbers are called *factors* of the larger number into which they all divide. For instance

$$60 = 2 \times 30$$
$$= 2 \times 2 \times 15$$
$$= 2 \times 2 \times 3 \times 5$$

Those in the last row are called *prime factors* because we cannot divide them further. By arranging these prime factors we can find the other numbers which will divide exactly into 60

$$2 \times 3 \times 2 \times 5$$
$$6 \times 10$$

Find as many other arrangements as you can.

> **Exercise 2.5**
> Think of hours and minutes and work the following examples:
> (a) $\frac{1}{4} + \frac{1}{60}$
> (b) $\frac{7}{12} + \frac{19}{60}$
> (c) $\frac{49}{60} - \frac{1}{2}$
> (d) $\frac{9}{10} + \frac{5}{6} + \frac{11}{60}$
> (e) $\frac{3}{20} + \frac{2}{3} + \frac{1}{2}$
> (f) $\frac{11}{30} + \frac{3}{4} + \frac{1}{3}$

Changing denominators to powers of 10

Modern methods of calculating depend on decimal fractions rather than common fractions. We must therefore find a way of changing

26 Calculating

our fractions so that they have denominators of 10, 100, 1000, etc.

To change fractions like $\frac{1}{2}, \frac{1}{3}, \frac{1}{4}, \frac{1}{5}$, etc., to decimals we first cut a whole one into ten pieces. The simplest to work out are $\frac{1}{2}$ and $\frac{1}{5}$, which are the same as $\frac{5}{10}$ and $\frac{2}{10}$ or 0·5 and 0·2.

Unfortunately 2 and 5 are the only numbers that are factors of 10. If we divide 10 by any other figure we will have a remainder. For example, to change $\frac{1}{4}$ into a decimal we first cut up our whole one into ten pieces which we divide by four much as we might share out ten apples among four boys. The result is two each but there are two over.

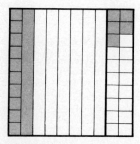

Figure 2.1

We now cut up each of the remaining two pieces into ten parts, as in Figure 2.1. We now have 20 much smaller pieces. They are in fact hundredths of the original square, so when we take $\frac{1}{4}$ of these we find we have $\frac{5}{100}$. So when we divide our square by 4 we have $\frac{2}{10}$ and $\frac{5}{100}$, or 0·25.

If we divide 1 by 8 following this system we first get $\frac{1}{10}$ and a remainder of $\frac{2}{10}$. The remainder is cut up as we did before so that we get 20 small pieces (hundredths). Dividing by 8 this time we get an answer of 2 hundredths but there are still 4 left. On our diagram these four squares are so small that we would find it impossible to divide them again into 10, but if we imagine this being done we will find we should have 40 thousandths. When divided by 8 this gives

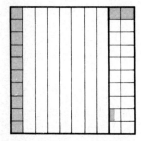

Figure 2.2

5 thousandths and no remainder. Since 5 thousandths is a half of 1 hundredth, we have shaded half a small square (see Figure 2.2). So $\frac{1}{8}$ is the same as $\frac{1}{10} + \frac{2}{100} + \frac{5}{1000}$ or 0·125.

We could arrive at the same result by simple division extended to fit our number line based on 10.

$$\begin{array}{r} 1 \quad \tfrac{1}{10} \quad \tfrac{1}{100} \quad \tfrac{1}{1000} \quad \tfrac{1}{10\,000} \end{array}$$

One-half $\quad 2\overline{)1 \cdot {}^1 0}$
$\, 0 \cdot 5$

One-quarter $\quad 4\overline{)1 \cdot {}^1 0 \; {}^2 0}$
$\, 0 \cdot 2 \; 5$

One-eighth $\quad 8\overline{)1 \cdot {}^1 0 \; {}^2 0 \; {}^4 0}$
$\, 0 \cdot 1 \; 2 \; 5$

Now work out for yourself $\frac{1}{9}$, $\frac{1}{3}$, $\frac{1}{6}$ and $\frac{1}{7}$. Each time you have a remainder put it in front of 0 in the next column and go on dividing. The results are more interesting, amusing and exasperating than the previous ones.

It seems that these fractions will not work out exactly. How exact do you want to be? Use the side of your ruler which has inches divided into eighths and rule a line $4\frac{1}{8}$ inches long. Now use the inches and tenths and try to rule a line exactly 4·125 inches long. When we write $4\frac{1}{8}$ inches we imply that the figure is correct to the nearest $\frac{1}{8}$ inch. When we write '4·125' we imply that it is correct to one-thousandth of an inch. To measure accurately 4·125 inches you would need some kind of micrometer. How accurate, do you think, was your original $4\frac{1}{8}$ inches?

In any practical situation we decide how exact we can expect to be and choose a suitable number of decimal places. We have now come upon a confusion in the way we use number. We may count separate objects — people, cars, screws or roses — and be positive that the number is correct. Less than whole ones of these would be of no practical use.

We also use numbers, not to count, but to measure, for instance, length, area, volume, weight, and so on. Each measurement is only approximate since the quantities are not disconnected units.

Calculating

Ratios

Factors of numbers will help us to simplify ratios by cancelling. Bill, using an old lathe, makes 32 components in a day. Jim, with a new machine, makes 48 similar components in a day. We can compare the two outputs with the ratio $32:48$ or $\frac{32}{48}$. If we replace the numbers with factors we have

$$\frac{2 \times 2 \times 2 \times 2 \times 2}{2 \times 2 \times 2 \times 2 \times 3}$$

By cancelling as we did in Chapter 1, we reduce this to $\frac{2}{3}$. It is often just as easy to cancel without finding factors first. The numbers 32 and 48 can each be divided by 8 and we can then cancel with 2.

$$\frac{\cancel{32}^{\cancel{4}^2}}{\cancel{48}_{\cancel{6}_3}} \qquad \text{So} \qquad \frac{\text{Bill's output}}{\text{Jim's output}} = \frac{2}{3}$$

Round figures

A great many of the measurements we use are only approximations. Consider the population of Great Britain. It is obviously impossible to give a figure for the exact population at any one moment in time. At intervals a census gives us the nearest figure that it is possible to calculate. During the night to which a census count applies there will be deaths and births, and some people always slip through the census without being counted.

The population figures are therefore given to the degree of accuracy that is suitable. Often only a very rough approximation is wanted. For instance, the population of Great Britain might be said to be 54 000 000. We call this a *round number* because it is rounded off, in this case to the nearest million.

The maps in this country in which scale is referred to as $\frac{1}{4}$ inch or 4 miles to an inch are, in fact, drawn in a ratio of $1:250\,000$, that is, 1 inch on the map represents 250 000 inches on the ground. This figure is as nearly accurate as it is possible to measure. If we took 1 inch to represent 1 mile the ratio would be $1:253\,440$. Thus when we say that a map is drawn on a scale of 4 miles to an inch, we are using a round figure.

A round figure provides only the degree of accuracy necessary for a given purpose. Incidentally it is also easier to remember.

Here is an example. A young Frenchman says: 'My girl friend's vital statistics are 86·36, 60·96, 81·28.'

Such a statement would be absurd. It is unlikely that any Frenchman would say it. Remember that he is measuring in centimetres. What would be the equivalent in inches if 1 inch = 2·54 centimetres?

The slide rule

Now we will take the simple slide rule (see Chapter 1, Figure 1.10) and see how to use it to divide. Move the top strip to the left so that 32 on the top strip is over 8 on the bottom strip (see Figure 2.3).

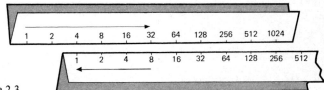

Figure 2.3

If you count the spaces between 1 and 32 you will find five. Between 1 and 8 on the bottom strip there are three. Thus the slide rule in this position shows five spaces minus three spaces, which leaves two spaces. The figure which is two spaces from the number 1 on the top scale is 4 which, as you already know, is the answer to 32 ÷ 8.

This corresponds to the division using powers of 2 which we learned earlier

$$\begin{array}{cc} & 32 \quad\quad 2^5 \\ \div & 8 \quad\quad 2^3 \\ \hline & 4 \quad\quad 2^2 \end{array}$$

Remember that the division sign, ÷, represents a fraction. Any division can be written to look like a fraction. A mathematician would be just as likely to read $\frac{32}{8}$ as 32 divided by 8 as he would to read it as 32 eighths.

Now look at all these pairs of numbers

```
4   8   16   32   64   128   256   512   1024
1   2    4    8   16    32    64   128    256
```

You will see that if you read them as divisions they all have the same answer, 4. This is found above the number 1, so we may read it as 4 whole ones.

Now try these on your simple slide rule

$$\frac{16}{2} \quad \frac{64}{4} \quad \frac{256}{128} \quad \frac{1024}{16} \quad \frac{512}{8} \quad \frac{128}{16}$$

The standard slide rule is numbered differently, but it is worthwhile setting $4 \div 1$ on it and comparing the answer with your findings on your simple slide rule. You may find out for yourself quite a lot about the markings on the slide rule which we will be discussing later.

Chapter 3

[Bank statement for A.N. OTHER ESQ., National Bank Limited, HIGH STREET, ANYTOWN]

Date	Detail		Debits	Credits	Balance when overdrawn marked DR
1971	BALANCE FORWARD				23.36 DR
20 APR		15334	8.00		31.36 DR
21 APR		15332	5.93		48.89 DR
		15335	11.60		51.82 DR
22 APR		15333	4.93		59.82 DR
28 APR		15336	6.00		
29 APR	CR. TRANSFER			52.58	
		15337	6.00		13.24 DR
30 APR		96973	4.50		17.74 DR
4 MAY	CR. TRANSFER			143.30	
		15339	5.00		
		15340	5.00		
5 MAY		15341	6.75		115.56
		15342	4.00		104.81
11 MAY		15344	4.00		100.81

Account No. 01615467 Statement No. 23

A.N. OTHER ESQ.,
16 HIGH STREET,
ANYTOWN,
AB1 2LD

DR on the bank statement means 'overdrawn'. That means that the customer owes money to the bank. The figures for the amount overdrawn used to be in red ink, from which comes the phrase 'in the red'.

Negative numbers

In a game of rummy a mathematician recorded his scores like this

		Running total
Won	10	10
Lost	15	−5
Lost	4	−9
Won	25	
Won	52	
Lost	30	
Lost	39	

Can you fill in the missing totals?

This is one example of a situation requiring numbers less than 0. These are called *negative numbers*.

Two people can play a simple game with two counters and two dice of different colours, say one blue and one red. A strip of paper is ruled and numbered as in Figure 3.1; it is large enough to allow the counters to be placed above and below the centre strip.

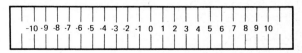

Figure 3.1

The players in turn throw both dice together. The blue gives the moves to the right (positive numbers) and the red gives moves to the left (negative numbers). Play begins with both counters on 0. The player moves his counter according to the combined score of each throw. If, for instance, his blue dice scores 6 and his red dice scores 4, his move will finish 2 to the right. If, on his second turn, his blue scores 2 and his red 5, his move will be 3 to the left because $2 - 5 = -3$. He will therefore go from 2 to -1.

If a player reaches 10 (on the right) he has won, and if a player reaches -10 (on the left) he has lost.

Powers

We have seen how numbers which can be written as powers of the same base number can easily be multiplied and divided.

Now look at this example. To find 16^3 (that is, $16 \times 16 \times 16$)

$$\begin{array}{rr} & 16 \\ \times & 16 \\ \times & 16 \\ \hline & 4096 \end{array} \qquad \begin{array}{r} 2^4 \\ 2^4 \\ 2^4 \\ \hline 2^{12} \end{array}$$

Answer: $16^3 = 4096$

A simpler method of writing the same calculation is

$$\frac{16^3}{4096} \qquad \frac{2^{4 \times 3}}{2^{12}}$$

Use this method to work Exercise 3.1.

Exercise 3.1
Find:
(a) 8^6
(b) 32^4
(c) 4^8
(d) 512^2
(e) 64^3

Your table of powers of 2 will allow you to work out the multiplications in Exercise 3.2. Here is an example of the working.

Find $8^2 \times 16^3$.

$$8^2 = 2^{3 \times 2}$$
$$\times\ 16^3 = 2^{4 \times 3}$$
$$\overline{262\,144 = 2^{18}}$$

Answer: $8^2 \times 16^3 = 262\,144$

Exercise 3.2

Find: (a) $2^7 \times 4^2$ (b) $32^2 \times 4^2$ (c) $256^2 \times 8$
(d) $64^3 \times 2^2$ (e) $128^2 \times 8^2$

Decimal fractions and degrees of accuracy

Anyone faced with the need to sum up a situation or an argument quickly (for example, in discussing a new piece-rate) cannot do complicated calculations. It is necessary sometimes to be able to say that a figure is 'about so much'. It is better if this kind of answer – an *estimation* – is limited, that is, the person making the estimation should know positively that it must be greater than one figure and less than another.

Look at this number line

0 10 20 30 40 50 60 70 80 90 100

Where would you put 43? It must come between 40 and 50. It is more than 40 and less than 50. We have signs to show this in order to save space. We take the sign = and turn it either to <, meaning 'is smaller than', or to >, meaning 'is greater than'.

We can now write $40 < 43$. The 40 is at the smaller end and so we read this as '40 is smaller than 43'. We could also write $43 < 50$ which means 43 is smaller than 50.

Putting these two together we would get

$$40 < 43 < 50$$

We can read this as: '43 lies between 40 and 50'.

Using these signs, show where the following numbers come in the number line

71   13   58   94   82   66

Consider a new number line

1·0   1·1   1·2   1·3   1·4   1·5   1·6   1·7   1·8   1·9   2·0

Here we have made ten divisions between 1 and 2.

34   Calculating

How would we place a number like 1·27? Remember that 1·2 means a whole one and two tenths. In 1·27 the 7 represents seven hundredths. We can see more clearly where to place the 1·27 if we put hundredths into our number line. Our number line now looks like this

1·00 1·10 1·20 1·30 1·40 1·50 1·60 1·70 1·80 1·90 2·00

We can now see that 1·27 lies between 1·20 and 1·30, or between 1·2 and 1·3.

Obviously 1·27 lies nearer to 1·30 than it does to 1·20, so to make an estimation we can now say that 1·27 is approximately 1·3, or 1·27 is 1·3 *correct to one place of decimals*.

We can write where a number would come in the number line and give the nearest approximate value that is correct to one place of decimals, in these ways

$$1·2 < 1·27 < 1·3$$

$$1·27 \approx 1·3$$

Notice the wavy 'equals' sign. This is a quick way of writing 'is approximately equal to'.

Exercise 3.3

Express the following correct to one place of decimals:

(a) 1·52
(b) 1·86
(c) 2·49
(d) 7·64
(e) 9·13
(f) 5·28

Go back to the number line we had above, namely

0 10 20 30 40 50 60 70 80 90 100

We can now place any number from 0 to 100 by imagining we have a camera with a zoom lens. Take the number 43·27. Look at one *digit* at a time. *'Digitus'* is the Latin for finger. We use the word digit to mean a single figure. The first digit, 4, represents 40 so we zoom up to the space between 40 and 50 and imagine this divided into 10 parts. The process is repeated for each digit in the number 43·27 as follows:

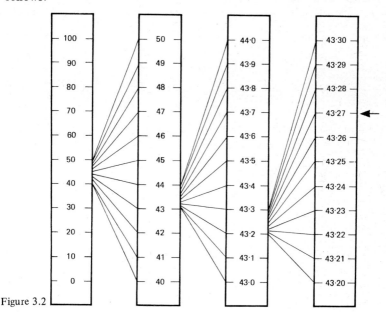

Figure 3.2

This process can in theory be continued indefinitely. In any practical problem we should first decide how accurate our number needs to be.

Recurring decimals

In Chapter 2 we suggested that you change to decimals the fractions $\frac{1}{9}$, $\frac{1}{3}$, $\frac{1}{6}$, $\frac{1}{7}$. You would have discovered when you were working the division that you always had a remainder. $\frac{1}{9}$ is 0·1111 .... We have a sign which means that the same number recurs. It is simply a dot over the figure. So we write $\frac{1}{9} = 0·\dot{1}$, and read this as 'nought point one recurring'. Similarly $\frac{1}{3} = 0·\dot{3}$.

When $\frac{1}{6}$ is divided out it becomes 0·16666 ... which we write 0·1$\dot{6}$.

But $\frac{1}{7}$ is different. When divided it becomes 0·142857142857 .... Here we can see a recurrence of six figures. This is written with two dots, one over the first and one over the last.

$$\frac{1}{7} = 0·\dot{1}4285\dot{7}$$

36   Calculating

When we decide how accurate we wish to be, we do not write down a recurring decimal, but write $\frac{1}{9} = 0\cdot 1$ or $0\cdot 11$ or $0\cdot 111$ depending on whether we choose one, two or three places of decimals respectively.

In the case of $\frac{1}{6}$ we must remember that $0\cdot 16$ is nearer to $0\cdot 2$ than it is to $0\cdot 1$, so we must say that $\frac{1}{6}$ is $0\cdot 2$ correct to one place of decimals. If we want two places of decimals we write $0\cdot 17$ because $0\cdot 166$ is nearer to $0\cdot 17$ than it is to $0\cdot 16$. The right way to write $\frac{1}{6}$ correct to three places of decimals would be $0\cdot 167$.

Similarly $\frac{1}{7}$ is $0\cdot 1$, $0\cdot 14$ or $0\cdot 143$. However, when we come to finding $\frac{1}{7}$ correct to four places of decimals we should notice that the fifth figure is 5, halfway between 1 and 10. It is generally accepted that we count 5 as we would 6, 7, 8 or 9, that is, we write $\frac{1}{7} = 0\cdot 1429$. Plan a diagram so that you can illustrate $0\cdot 999\ldots$ to as many decimal places as possible. It will be similar in some ways to Figure 1.1 in Chapter 1 which illustrated $\frac{1}{2}, \frac{1}{4}, \frac{1}{8}$, etc.

Ratios as decimals

We have seen ratios set down as fractions. A ratio of $2:5$ can be written as $\frac{2}{5}$. It can also be changed to a decimal fraction by dividing $2\cdot 0$ by 5. The result is $0\cdot 4$; this form is the most convenient when using a slide rule or calculator.

Remember to decide how accurate you wish your decimal equivalent to be. In the case of $\frac{5}{6}$, after dividing $5\cdot 0$ by 6 we have $\cdot 8333$. The 3 is in fact recurring so for our final figure we simply choose the number of places of decimals that we want.

To be correct to, say, three places of decimals we must work out four places and then take the nearest approximation to three places, as we saw when finding the decimal equivalent of $\frac{1}{6}$. For example

$$\frac{2}{7} = 0\cdot 2857$$

$$= 0\cdot 286 \text{ correct to three places of decimals}$$

Exercise 3.4

Find, correct to 3 places, the decimal equivalents of these fractions:

(a) $\frac{1}{7}$ (d) $\frac{5}{7}$
(b) $\frac{3}{7}$ (e) $\frac{6}{7}$
(c) $\frac{4}{7}$

The decimal point

1p × 100 = £1. This may be written another way: £0·01 × 100 = £1·00. The 1 in £0·01 has moved two steps along to make the amount of money larger. The same is true for any amount of money.

For example

$$£0·05 \times 100 = £5·00$$
$$£0·24 \times 100 = £24·00$$
$$£7·00 \times 100 = £700·00$$
$$£7·24 \times 100 = £724·00$$

In the same way any quantity expressed as a decimal can easily be multiplied by 100. Here are some examples

$$1·26 \times 100 = 126·00$$
$$4·792 \times 100 = 479·2$$
$$8·1 \times 100 = 810·0$$
$$1·9302 \times 100 = 193·02$$

Exercise 3.5

Multiply by 100:

(a) 7
(b) 300
(c) 3·5
(d) 210
(e) 2·1
(f) 0·37
(g) 5·821
(h) 6·424

Estimate first.
  For example, (h) is 6 and a fraction of 6. When multiplied by 100 it should be 600 and something.

Exercise 3.6

Multiply by 10:

(a) 3·4   (d) 6·041
(b) 8·12 (e) 201·8
(c) 42·3 (f) 300·9

Estimate first.
  For example, in (f) 300 × 10 is 3000, so the answer should be 3000 and something.

38 Calculating

Dividing by 10 and 100 can be just as easy if we first make an estimate by looking at the whole numbers. For instance, 32 divided by 10 equals 3 and a bit so 32 ÷ 10 is 3·2.

Exercise 3.7

Divide by 10:   Divide by 100:
(a) 15   (f) £1
(b) 27·4   (g) £5
(c) 81   (h) £37
(d) 128   (i) 137
(e) 1024   (j) 120·8

There are two important things to remember:

1. **The decimal point marks the end of the whole number.** If a number is written without a decimal point it can always be put in. For example, 15 is the same as 15·0.

2. **Multiplying by whole numbers makes numbers larger; dividing by whole numbers makes them smaller.**

We have shown how the figures move across the decimal point when multiplying and dividing by 10 and 100. The same rules apply with larger powers of 10, that is, for example, when multiplying by 1000 the figures move across the decimal point three places to make a larger number. Dividing by 10 000, the figures move across the decimal point four places to make a smaller number.

The same system works for any power of 10: the number of places moved is the number of zeros where we have multipliers in the form 100, 1000, 10 000, etc. Sometimes these are written as powers of 10, that is 10^2, 10^3, 10^4. The indices show the number of zeros in each number.

Anyone who works regularly with figures of this kind comes to think of the process as 'moving the decimal point'. This can be very confusing for those still learning to understand decimals. It is important to remember that it is the value of the number which changes.

Computer numbers

It is interesting to compare the multiplication and division of numbers by 10 or 100 with binary numbers when multiplied or divided by powers of 2 written in base 2. In the following examples the ordinary (base 10) numbers are shown in brackets

Base 2		Base 2		Base 2
11 (3)	×	10 (2)	=	110 (6)
1 010 (10)	÷	10 (2)	=	101 (5)

Exercise 3.8

Fill in the missing numbers (base 10 numbers in brackets):

	Base 2		Base 2		Base 2
(a)	1 (1)	×	10 (2)	=	10 ( )
(b)	10 (2)	×	100 (4)	=	1000 ( )
(c)	11 (3)	×	100 (4)	=	( )
(d)	111 ( )	×	10 ( )	=	( )
(e)	101 ( )	×	10 ( )	=	( )
(f)	100 (4)	÷	10 (2)	=	( )
(g)	10 110 (22)	÷	10 (2)	=	( )
(h)	110 ( )	÷	10 (2)	=	( )
(i)	1 000 ( )	÷	100 ( )	=	( )
(j)	11 100 ( )	÷	100 ( )	=	( )

Percentages

We do not need a formula for calculating percentages. Here are three simple steps by which to find 50% of £3

$$1\% \text{ of } £1 = 1p$$
$$50\% \text{ of } £1 = 50p$$
$$50\% \text{ of } £3 = 50p \times 3 = 150p = £1 \cdot 50$$

To find 6% of 50p we have to remember that 50p is half of £1, so we may write

$$1\% \text{ of } £1 = 1p$$
$$1\% \text{ of } 50p = \tfrac{1}{2}p$$
$$6\% \text{ of } 50p = \tfrac{1}{2}p \times 6 = 3p$$

Now suppose you need to find 18% of £236·50. Work out the

percentages step by step and then total them in the following way

18% of £200 =
18% of £30 =
18% of £6 =
18% of £0·50 = _____
18% of £236·50 = _____

This is a useful method of working complicated figures and one which does not require a long multiplication sum, which would be as follows

1% of £236·50 = $\frac{1}{100}$ of £236·50 = £2·365

18% of £236·50 = 18 × £2·365

```
  2·365
     18
 ──────
 18·920
  23·65
 ──────
 42·570
```

Answer: 18% of £236·50 = £42·57

Exercise 3.9

Calculate:

(a) 9% of £152·50
(b) 20% of £348·25
(c) 12% of £604·75
(d) 18% of £215·50
(e) Out of a consignment of 250 tons of ore, 60% of metal is recovered. What will be the weight of metal produced?

Choose the easiest method. For example, 9% is the same as 10% less 1%.

We started this section with 1% of £1 = 1p; or 1% of £1 = £0·01. A similar statement would also be true of metres: 1% of 1 metre = 1 centimetre; or 1% of 1 metre = 0·01 metres. So we can see that, in fact, no matter what units we use, 1% of 1 = 0·01.

From this, a percentage increase can be calculated

$$1\% \text{ of your wage} = 0\cdot01 \times \text{your wage}$$

$$7\% \text{ of your wage} = 0\cdot07 \times \text{your wage}$$

$0\cdot07$ becomes your *rate of increase*.

Your total pay after receiving a rise is easily found by adding the old wage and the rise together. These two calculations, finding the rise and adding it to the original wage, can be done in one calculation. If we write w for weekly wage and r for the rate of increase we can make up this formula

$$w \times 1 + w \times r$$

or more simply $w(1 + r)$

Now if r is $0\cdot07$ and is added to 1 we get $1\cdot07$. So your new wage can be found by multiplying your old wage by $1\cdot07$.

Shopkeepers buy at one price and sell at another. For example, sugar costs 10p; the mark up for selling is 5%. In this case the selling price would be $10p \times 1\cdot05$ which equals $10\cdot5p$.

Exercise 3.10

(a) At a coal mine it is stated that productivity (that is, output per man shift in this case) has gone up by 4%. If the output is approximately 5000 tons a week, how much coal will be produced per week following the increase in productivity?

(b) The cost of living is stated by the Government to have gone up by 8%. If a worker's net take-home pay is £30 per week, how big will his take-home pay need to be to compensate him for the increase in the cost of living?

(c) The Gross Domestic Product (that is the total value of all goods and services produced) of the UK in 1972 prices was £52 667 million. Assuming no change in prices, what would the GDP be in 1973 assuming that it grew by 5%?

Chapter 4

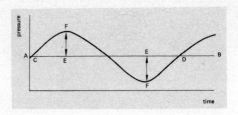

This chapter introduces some very simple graphs. These present information in a form which gives a picture of a situation. It is much more easily and quickly understood than the information would be if set out in words and figures. An understanding of graphs is essential for a study of statistics.

Powers of 2

In Chapter 3 we learned how to multiply a number by itself using powers of 2. For example: find 64^3.

Long form		Short form
64	2^6	$64^3 = 2^{6 \times 3}$
× 64	2^6	
× 64	2^6	
262 144	2^{18}	$262\,144 = 2^{18}$

Answer: $64^3 = 262\,144$

Suppose now that you were asked the reverse: what number has 1 been multiplied by 3 times to make 262 144? (Do not forget that all multiplications start from 1.) We will begin with 262 144 and get to the answer at the end of our working. To help us here we need one of those mathematical signs that save us from writing a lot of words. In this case the sign is $\sqrt[3]{}$ (the cube *root* of) instead of 'the number that 1 has been multiplied by 3 times to make'.

Using the table of powers of 2 (see Chapter 1), the calculation can now be written like this:

$$262144 = 2^{18}$$
$$\sqrt[3]{262144} = 2^{18} \div 3$$
$$64 = 2^6$$

Answer: $\sqrt[3]{262\,144}$ is 64

Compare this with the short form used to find 64^3 and you will see how the process has been reversed.

You will sometimes see the sign $\sqrt{\ }$ and no number. This always means the square root of .... The length of the side of any square is the root of the square.

> **Exercise 4.1**
> Find:
> (a) $\sqrt[3]{32\,768}$
> (b) $\sqrt[5]{32\,768}$
> (c) $\sqrt[4]{4096}$
> (d) $\sqrt[6]{4096}$
> (e) $\sqrt[9]{262\,144}$
> (f) $\sqrt{256}$
> (g) $\sqrt{65\,536}$
> (h) $\sqrt{64}$
> (i) $\sqrt{16}$
> (j) $\sqrt{4}$

Binary number

In Chapter 2 we saw how the powers of 2 are used for computers. You will remember we had the headings

$$64 \quad 32 \quad 16 \quad 8 \quad 4 \quad 2 \quad 1$$

Under these we wrote either 1 or 0.

Look at this binary number: 1 001 101. Without the headings it is very difficult to see at a glance how many this represents.

In ordinary numbers we used to put commas to group our figures in sets of three, for instance 1,325,684. This made it easy to read. However, since on the Continent the comma is used as a decimal point, we now drop the comma and leave a space: 1 325 684. It is just as easy to see that this is one million, three hundred and twenty-five thousand, and so on.

Notice that we read each block of three figures as though they are hundreds, tens and units, and then add the word thousand or million as the case may be.

We may, if we find it easier, read binary number in much the

44 Calculating

same way. The second set of three figures from the right in our example 1 001 101, which would be thousands in decimal number, is eights in binary number.

In 1 001 101 the 001 means 1 eight. Similarly the 1 on the far left, which would have been 1 million in decimal number, is now 1 sixty-four. The total is $1 \times 64 + 1 \times 8 + 5$, that is, 77.

We have only to recognize possible arrangements of the first set of three figures (that is, the set on the right) which takes us from 1 to 7, and a six-figure binary number will be easy to read.

Use the table on the left to convert the numbers in Exercise 4.2 to decimal numbers as in the example at the top.

Binary

001 = 1
010 = 2
011 = 3
100 = 4
101 = 5
110 = 6
111 = 7

Exercise 4.2
Example: $011\ 100 = 3 \times 8 + 4 = 28$
(a) 101 010
(b) 111 111
(c) 100 001
(d) 1 110 110
(e) 1 001 101
(f) 1 010 011

The opposite process is used to change ordinary numbers to binary numbers.

Here is an example. Take 57. First divide by 8. In this case there are 7 eights and 1 more. The number 7 is written 111 and 1 is 001, so 57 is written 111 001.

Exercise 4.3
Change to binary numbers:
(a) 43 (b) 29 (c) 17 (d) 63 (e) 39
(f) 11 (g) 50 (h) 15 (i) 31 (j) 23

Fractions

Figure 4.1 will help you to understand the multiplication of fractions.

This represents a shoe factory where a quarter of the space is taken up by packaging. A third of the space is used in the production of children's shoes. How much space is used packaging children's shoes?

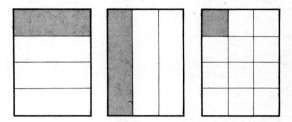

Figure 4.1

The figure shows what happens when we find $\frac{1}{3}$ of $\frac{1}{4}$. In the rectangle on the far right there are twelve equal parts, so the shaded one is $\frac{1}{12}$. We have multiplied two top numbers of the fractions: $1 \times 1 = 1$. We have also multiplied the two bottom numbers of the fraction: $4 \times 3 = 12$.

Make a similar diagram, but shade $\frac{3}{4}$ and $\frac{2}{3}$. Illustrate the result when these are multiplied. Is it the same as it would be if you took 3×2 for the top figure and 4×3 for the bottom figure? You should simplify the answer by cancelling where possible. Check the final answer with your diagram.

> **Exercise 4.4**
> Find:
> (a) $\frac{2}{5} \times \frac{3}{4}$ (d) $\frac{11}{12} \times \frac{4}{5} \times \frac{1}{11}$
> (b) $\frac{1}{2} \times \frac{5}{7}$ (e) $\frac{8}{9} \times \frac{3}{4} \times \frac{3}{8}$
> (c) $\frac{5}{9} \times \frac{3}{10}$ (f) $\frac{5}{12} \times \frac{3}{10} \times \frac{8}{9}$

Decimal fractions

Multiply 0·1 by 0·1. This is a favourite catch question of school examinations. Before you decide on your answer think how you could draw a diagram similar to that in Figure 4.1.

As soon as you have to visualize 0·1 you must think of $\frac{1}{10}$. If you now multiply $\frac{1}{10}$ by $\frac{1}{10}$ you will see why the answer is $\frac{1}{100}$. So 0·1 × 0·1 is 0·01. Each factor in the multiplication had one place of

46 Calculating

decimals and the *product* (the result of the multiplication) has two.

Now consider 0.01×0.01. This means $\frac{1}{100} \times \frac{1}{100}$ so we can see that the answer will be $\frac{1}{10\,000}$. In the decimal system this is 0.0001, that is, $0.01 \times 0.01 = 0.0001$.

Take the example 0.4×0.7. If you remember that this means $\frac{4}{10} \times \frac{7}{10}$, that is $\frac{28}{100}$, you will see that the answer is 0.28.

In these examples notice that the answers have a zero to the left of the decimal point, that is, all numbers are less than 1. This is logical when you consider that if you take a fraction of a fraction the result must be smaller still. If you look back to Figure 4.1 again you will notice that the result of multiplying $\frac{1}{4} \times \frac{1}{3}$ is a comparatively small piece.

Multiplying by a fraction makes numbers smaller.

Exercise 4.5

Multiply:
(a) 0.6×0.9
(b) 0.8×0.2
(c) 0.3×0.1
(d) 0.7×0.4
(e) 0.5×0.5
(f) 0.9×0.9

Graphs

Where a proportion exists, a graph may be drawn first to represent the proportion and then to read off information. For example, the output of a plant is 50 tons per hour. In 2 hours the output would be 100 tons; in 3 hours 150 tons and so on.

In Figure 4.2 we have vertical lines which can represent the tons produced in a certain number of hours.

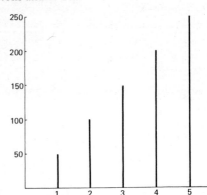

Figure 4.2

Now take an example of a different kind. Bloggs & Cogs Engineering Ltd. are hit by a 'flu epidemic. On Monday 50 workers are absent; on Tuesday 100; on Wednesday 150; on Thursday 200 and on Friday 250. These absences could also be represented by the vertical lines on Figure 4.2.

A 'flu epidemic, however, is unlikely to strike with such a mathematical regularity. We do not expect to see the number of 'flu victims being exactly proportional to the passage of time. When considering an output of 50 tons per hour we could draw another vertical line between 2 and 3 to show that in $2\frac{1}{2}$ hours the output would be 125 tons. We could also extend the diagram to find the output in 6, 7 or 8 hours. In the case of the 'flu epidemic what would such lines show? Between 2 and 3 would mean between Tuesday and Wednesday, that is, during the middle of the night. But Bloggs and Cogs don't have a night shift, so this would be meaningless. In the same way the numbers 6 and 7 would represent Saturday and Sunday when again there are no workers in the factory. And when we come to 8 we are representing the following Monday morning. There will be workers in the factory but we can draw no conclusion from the previous week, in spite of its apparent regularity, to help us to estimate the number of absentees on this Monday.

These two sets of figures need entirely different treatment.

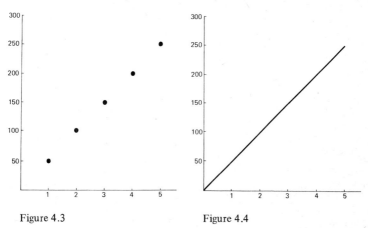

Figure 4.3   Figure 4.4

In Figure 4.3 we have the same information as in Figure 4.2 but here the output is marked by crosses which show the number of tons produced in 1 hour, 2 hours, and so on. Now since we have already agreed that we could also calculate an output of $2\frac{1}{2}$ hours we can see that we could also add the output in $\frac{1}{2}$, $1\frac{1}{2}$, $3\frac{1}{2}$ or $4\frac{1}{2}$ hours. Indeed we could include $\frac{1}{4}$, $2\frac{1}{8}$, $3\frac{1}{16}$, $4\frac{11}{32}$, and so on. There is no end to the number of points we could put in the graph. In mathematical language we have an *infinite* number of points in this set of points. Figure 4.4 shows this infinite number of points. There are no spaces between the points so it becomes a straight line.

Now look at Figure 4.5. Here we have a bar chart. It shows the number of workers absent each working day. It would be wrong to join the tops of these bars with a straight line because as we have seen they are isolated cases. The line graph and the bar chart must be 'read' in quite different ways.

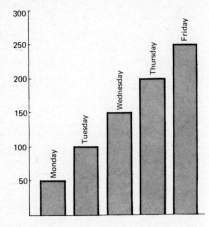

Figure 4.5

Significant figures

The distance round the earth (great circle) is about 25 000 miles. The distance of the moon from the earth is about 240 000 miles. 1 in ≈ 2·5 cm. 1 ft^2 ≈ 0·093 m^2. The first two of these figures we call 'round figures'. The figure in centimetres is correct to one place of decimals, and the figure in square metres is correct to three places of decimals.

Can you see one feature which is common to all four?

Leaving aside the zeros, each of these numbers has two digits. These are the two figures which are significant. The zeros and the decimal point are necessary only to show where these two digits appear on the decimal number line. Thus all four examples are said to be 'correct to two *significant figures*'.

The commercial slide rule

The slide rules which can be bought come in different styles. The basic principle is the same for them all, but some are fixed up for financiers, others for engineers, and so on. Any trade which is concerned with the multiplication and division of numbers can use a slide rule.

The big thing to remember is that the slide rule always gives an approximate answer to a problem, an answer which is said to be correct only to two or three significant figures.

Probably the most common basic design is that shown in Figure 4.6.

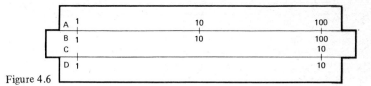

Figure 4.6

If you look at such a slide rule you will see that the diagram has left out all the graduations in between 1 and 10 and between 10 and 100. Also you are unlikely to find the scales lettered A, B, C and D as in the diagram. The lettering is merely to help us to identify them.

You will find that the sliding part of the rule (which carries the B and C scales) runs in grooves between the A and D scales. There will probably also be a Perspex window, the *cursor* (Latin for 'runner'), which can be moved along the rule. It has a fine vertical line, the hairline, which we will use presently.

Take off the cursor and draw out the sliding rule carrying the B and C scales. You can now examine the A scale more clearly. You will see that on the left the printed numbers start at 1 and go on across to the right: 2, 3, 4, 5, 6, 7, 8, 9, 10, 20, 30, 40, 50, 60, 70, 80, 90, 100.

Mark the distance between 1 and 2 on a strip of paper. With this gauge measure the distance between 2 and 4 and between 4 and 8. What do you find?

You will see that if we add the distance from 1 to 2 marked on the paper gauge to the distance from 1 to 4 on Scale A, the mark on the gauge comes opposite 8. We have multiplied 2 by 4.

These are powers of 2. You can see that they can be multiplied on the standard slide rule as they were on the simple paper slide rule.

Will the same system work for other numbers than those in the table of powers of 2? Try, using a paper gauge.

Replace the sliding part of the rule so that 1 on the B scale is under 1 on the A scale. If you look along the slide rule now (examining only the A and B scales) you will see that one is an exact duplicate of the other.

Look at the space between 10 and 20 on the A and B scales. Can you recognize where 15 must be by the graduation which sticks up on the A scale (and down on the B scale) rather more than halfway between 10 and 20? If you can find 15 you should be able to identify the other main graduations which project a little but not as much as that for 15. If you count them along you will find they represent

11, 12, 13, 14,   16, 17, 18, 19

In order to be sure what a graduation represents it is always safer to count them along in this way. Between 10 and 100 you will find 20, 30, 40, 50, 60, 70, 80, 90.

Because the distances between numbers diminish as you go to the right of the slide rule there is less space for the graduations at the right-hand end. If you look at the A and B scales you will see that there are several systems of graduations, namely:

The graduations between 1 and 2 and between 10 and 20 are of the same pattern.
The graduations between 2 and 5 and between 20 and 50 are of the same pattern.
The graduations between 5 and 10 and 50 and 100 are of the same pattern.

Replace the cursor and move it so that the hairline marks in turn each of the following

1·5	1·3	1·9	15	13	19
2·5	3·35	4·95	25	33·5	49·5
5·1	6·8	7·2	51	68	72

As you move along the rule you must watch out for the changing system of graduations.

The cursor has two uses:

1. To line up numbers accurately.

Example: 16 × 2·5. Move the cursor so that the hairline is over 16 on the A scale. Move the slide to bring the 1 on the B scale below the 16, already marked by the hairline.

2. As a marker to read off answers.

Leave the slide exactly in position and move the cursor so that the hairline is over 2·5 on the slide (the B scale). The answer lies under the hairline on the A scale. It is 40.

Exercise 4.6

Calculate the following using the slide rule:

(a) 3×20 (f) $1 \cdot 5 \times 40$
(b) 15×4 (g) $1 \cdot 8 \times 5$
(c) 7×13 (h) $3 \cdot 5 \times 6$
(d) $2 \cdot 5 \times 6$ (i) $4 \cdot 2 \times 15$
(e) $1 \cdot 5 \times 20$ (j) $1 \cdot 24 \times 25$

Chapter 5

Mathematicians and scientists need a shorthand to summarize their discoveries and operations. This saves space and time. It may be a formula like $W = v \times a$, meaning, multiply volts by amps to get watts.

The theory of indices

We have discovered four rules for using the indices when numbers can be written as powers of 2:

Rule 1: To multiply: add.
Rule 2: To divide: subtract.
Rule 3: To find a power: multiply.
Rule 4: To find a root: divide.

This very short form of words does not explain *what* we add, subtract, etc. We can write the same rules more easily and yet say exactly what we mean if we use sign language.

Rule 1: $2^x \times 2^y \equiv 2^{x+y}$

x can be replaced by any number so long as the same number is put in the place of x on both sides of the sign $\equiv$. The same is true for y. The three lines used here mean 'equals' for all values of x and y. The other rules can be written

Rule 2: $2^x \div 2^y \equiv 2^{x-y}$

Rule 3: $(2^x)^y \equiv 2^{x \times y}$

Rule 4: $\sqrt[y]{2^x} \equiv 2^{x \div y}$ or $2^{\frac{x}{y}}$

Chapter 5 53

The table of powers of 2 can now be extended as follows

32	16	8	4	2	1	$\frac{1}{2}$	$\frac{1}{4}$	$\frac{1}{8}$	$\frac{1}{16}$	$\frac{1}{32}$
2^5	2^4	2^3	2^2	2^1	2^0					

If you look at the number pattern made by the indices you can put in the powers of 2 for the fractions. If you feel that you have simply made a guess don't worry. That is the way mathematicians often start.

You can find out if your guess is correct by working certain divisions using powers of 2. Remember that to divide we subtract the indices. Can you think of any division, using powers of 2, that ought to result in the answer $\frac{1}{2}$? Set this down as we did in Chapter 2 when first learning how to divide using indices. Does the result fit in with your guess? There is a clue in one of the subheadings in Chapter 3.

Here is another example.

$$\begin{array}{cc} & 8 \quad\quad 2^3 \\ \div & 32 \quad\quad 2^5 \\ \hline & \underline{} \quad\quad \underline{} \end{array}$$

Following the rule for division we subtract the indices: $3 - 5$ is -2 (see Chapter 3, page 31, on negative numbers). We can write $8 \div 32$ as $\frac{8}{32}$ which is the same as $\frac{1}{4}$.

$$\begin{array}{cc} & 8 \quad\quad 2^3 \\ \div & 32 \quad\quad 2^5 \\ \hline & \underline{\frac{1}{4}} \quad\quad \underline{2^{-2}} \end{array}$$

So 2^{-2} is the same as $\frac{1}{4}$ or $\dfrac{1}{2^2}$.

If we now look at the number line written across the page we can see that for figures below 2^0 we should have *negative indices*, namely

54 Calculating

$\frac{1}{2} \quad 2^{-1}$
$\frac{1}{4} \quad 2^{-2}$
$\frac{1}{8} \quad 2^{-3}$
$\frac{1}{16} \quad 2^{-4}$
$\frac{1}{32} \quad 2^{-5}$

Exercise 5.1

In (a)–(e) find n; in (f)–(j) give the fraction:

(a) $\frac{1}{8192} = 2^n$ (f) $2^{-14} =$
(b) $\frac{1}{1\,048\,576} = 2^n$ (g) $2^{-19} =$
(c) $\frac{1}{65\,536} = 2^n$ (h) $2^{-12} =$
(d) $\frac{1}{128} = 2^n$ (i) $2^{-8} =$
(e) $\frac{1}{2048} = 2^n$ (j) $2^{-6} =$

Ratio and proportion

In Chapter 1 we explained a ratio by comparing two gear wheels, one with a diameter of 9 inches and the other with a diameter of 3 inches. The ratio of the two wheels was 9:3 which cancelled to 3:1, or could be expressed as a fraction $\frac{3}{1}$. We might say that **a ratio is the number of times one quantity is contained in another quantity of the same kind.**

The ratio of the gear wheels was $\frac{3}{1}$ and not $\frac{9''}{3''}$. The comparison of the two gear wheels was 3 to 1. This ratio will be correct even if in fact both wheels are two or three times as large. If we want to have two gear wheels in the ratio of $\frac{3}{1}$ but of sizes other than 9'' and 3'', we will have to ensure only that one is three times larger than the other.

An expression such as $\frac{84 \text{ miles}}{4 \text{ hours}}$ is not a ratio because we are comparing different things. This expression means a speed of 84 miles in 4 hours, which we can cancel to $\frac{21 \text{ miles}}{1 \text{ hour}}$ or 21 miles per hour.

Assuming that the vehicle concerned maintains a constant speed we can say that the distance travelled will be proportional to the time taken.

In a proportion there are two equal ratios. Each ratio involves two quantities, so that there are four quantities altogether. In problems involving proportion we know three quantities and need to find the fourth.

In 4 hours a factory printed, glued and folded 48 000 boxes. How many boxes could be produced in 5 hours?

You may be able to do this in your head, but can you remember the steps by which you work it out? These steps form the basis of the system used for working more complicated examples on a calculator.

You probably divided by 4 to find the production for 1 hour and then multiplied by 5 to find the result for 5 hours. If we put this as one calculation we get $\frac{48\,000}{4} \times 5$. But the order in which the multiplying and dividing is done does not matter. We could have written $48\,000 \times \frac{5}{4}$. In this latter form we see $\frac{5}{4}$ as the ratio between the hours spent in production. The boxes produced are proportional to the hours spent, so the 48 000 boxes can be multiplied by the ratio $\frac{5}{4}$ which can be written as 1·25 for the calculator. The number of boxes produced in 5 hours is

$$48\,000 \times 1\cdot 25 = 60\,000$$

We can also solve problems of this kind by putting down the 2 pairs of figures that are in the same proportion. The production rate is agreed to be constant. First put 'b' for the figure we are asked to find. The proportions can now be written: $\dfrac{b}{5} = \dfrac{48\,000}{4}$. If the 2 fractions are equal, they will still be equal if we multiply each by the same number. In this case, multiplying by 5 we get

$$\frac{b \times 5}{5} = \frac{48\,000 \times 5}{4}$$

We can now cancel the fives on the left-hand side. This leaves

$$b = \frac{48\,000 \times 5}{4} = 60\,000$$

We can take a short cut by going straight from $\dfrac{b}{5} = \dfrac{48\,000}{4}$ to $\dfrac{b}{5} = \dfrac{48\,000}{4}$ to $b = \dfrac{48\,000 \times 5}{4}$. We have, in fact, chosen to multiply by 5 precisely because it gets rid of the figure under the 'b'.

Exercise 5.2

(a) How far will a car go on 6 gallons of petrol if it will go 54 miles on 2 gallons?

(b) A colliery produces 25 000 tons in 5 shifts. What should be the output in 8 shifts?

(c) How far will a car travel in 1 hour if it can do 2 miles in 3 minutes? (Notice the difference in the units: change 1 hour into its equivalent in minutes.)

(d) How many sacks of potatoes can be bought with £18 if 3 sacks cost 90p?

(e) How much would it cost to widen 3 miles of road if it costs £649 to widen 3 furlongs? (8 furlongs = 1 mile)

Estimations

Whenever we use an aid to calculation we must first know roughly how big the answer must be.

Take the following: find the total of 39 hours at 85p per hour. Now $39 \approx 40$ and $85 \approx 90$, so we first find what 40 hours at 90p makes. Take $4 \times 9 = 36$ and then add two zeros to make 3600p or £36. When we add two zeros what we are in effect doing is multiplying by 10 twice, since what we are multiplying is really $4 \times 10 \times 9 \times 10$.

Here is another example, writing down the steps which you can do in your head: 48 coach fares at £3·97 each.

Estimate: 50 coach fares at £4·00 each.

$$4 \times 5 \times 10 = 20 \times 10 = £200$$

Exercise 5.3

Estimate the following:

(a) Petrol: 11 gallons at $37\frac{1}{2}$p per gallon.
(b) Electricity: 2920 units at 0·92p per unit.
(c) Oil: 401 gallons at 11·9p per gallon.
(d) Rates: £186 rateable value; 32p in the £.
(e) Television rental: 26 weeks at 95p per week.
(f) Cat food: 72 tins at $6\frac{1}{2}$p per tin.

Formulae

Here is a table showing speeds, times and distances travelled:

50 mi/h	7 h	350 mi
64 ft/s	10 s	640 ft
100 km/h	$3\frac{1}{4}$ h	325 km

What would be the distance travelled if the speed were s mph and the time t hours?

You will have discovered that the speed is multiplied by the time to find the distance travelled. If we call the distance 'd' we can write this formula: $d = s \times t$. We know that this will be true for any speed and any time. This is the formula that connects distance, speed and time.

There is only one condition. The *units* must be the same. That is, if the speed is in miles per hour we must put the time in hours, or fractions of an hour, and the distance will be in miles. If the speed is in feet per second then the time must be given in seconds (1 minute would have to be written as 60 seconds and 1 hour as 3600 seconds), and the resulting distance will be in feet.

If you are paid by the hour at a rate that is the same for all the hours you work, the formula for your weekly wage will be similar to the distance/speed/time formula.

Make up your own formula for finding a simple weekly wage. Your own wage may be complicated by other considerations such as overtime. Can you make up a formula which will help you to check what your total pay should be?

Square roots

Figure 5.1

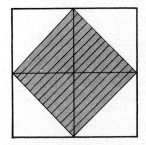

Figure 5.1 shows a square of 4 square inches. How long is each side? Obviously 2 inches. We use the term 'square root' in this situation. As we saw in Chapter 4, we have a sign for 'square root', so we could write simply $\sqrt{4} = 2$. We read this as 'The square root of 4 is 2.'

58 Calculating

> **Exercise 5.4**
> Find:
> (a) $\sqrt{16}$ (b) $\sqrt{9}$ (c) $\sqrt{25}$ (d) $\sqrt{100}$ (e) $\sqrt{10\,000}$
> Of what numbers are these the square roots?
> (f) 8   (g) 11   (h) 15   (i) 7   (j) 12

In Figure 5.1 we have a shaded area. Is this a square? What is its area?

The shaded area is a square and its area is two square inches (each triangle in the shaded area is half a square inch). The length of one side of this square must therefore be the square root of 2. If we measure it in inches and tenths of inches we find that it is about 1·4 inches long. If this is correct the area will be 1·4 in × 1·4 in = 1·96 sq in. We must have measured slightly short, but we are only 0·04 sq in out. We could hardly expect to be more accurate when working at this scale.

Try making an accurate drawing for yourself, but make it bigger. One decimetre would be a useful unit to replace the inch. A centimetre is a tenth of a decimetre and a millimetre is a hundredth of a decimetre, so you should be able to measure accurately to two places of decimals.

You may simply try guessing what the second place of decimals would be.

If we take the figure of 1·4 which we have already found and call the unit decimetres instead of inches, 1·4 will mean 1 decimetre and 4 centimetres. If you try adding on to this 1 millimetre, you will get 1·41 decimetres. Multiply 1·41 by 1·41. See how near to 2 this answer is. Then try to multiply 1·42 by 1·42. Which figure is nearer to 2?

In fact, as we have seen before, if we multiply two fractions (common or decimal) we will get an even smaller fraction, so we will never get an exact value for $\sqrt{2}$, although we know that it is possible to have a square with an area of two square inches or two square decimetres.

How absurd! The square root of 2 is in fact known as a *surd*, from the Latin '*surdus*' meaning mathematically irrational.

The slide rule

Take a standard slide rule, using the A and B scales. Slide the moving part so that the 5 is under the 1. You can easily pick out the following figures

1	2	3	4	5	6	7	8	9	10
5	10		20		30		40		50

In the last chapter we studied these markings, so you should have no difficulty now in seeing that 3 is over 15, 5 over 25, and so on. All these pairs of figures are in the same ratio to one another

$$\frac{1}{5} = \frac{4}{20} = \frac{7}{35} = \frac{10}{50} \text{ and so on}$$

We have already seen that the markings on the A and B scales are the same between 10 and 100 as they are between 1 and 10. If we had a longer rule we could repeat the same markings and number them from 100 to 1000 and again from 1000 to 10 000.

There is no need for such a long slide rule since the markings are the same. We simply read all the numbers as though they were 10 times or 100 times bigger as required.

Suppose we multiply all the bottom figures by 10. We would have

$$\frac{1}{50} \quad \frac{2}{100} \quad \frac{3}{150} \quad \frac{4}{200} \quad \frac{5}{250}$$

These are the same figures as the ones we had in our output graph in Chapter 4. You will notice that the proportions are still equal, that is

$$\frac{1 \text{ hour}}{50 \text{ tons}} = \frac{2 \text{ hours}}{100 \text{ tons}} = \frac{5 \text{ hours}}{250 \text{ tons}}$$

All these represent outputs of 50 tons per hour. We could find from our slide rule the output in any number of hours or parts of an hour. We could also find how many hours would be required for a given output.

Suppose we want to know how many hours it would take to produce 400 tons. Keeping the same setting, we slide our cursor to 40 on the B scale (remember that we are multiplying all the B scale numbers by 10, so 40 represents 400). The number marked by the cursor on the A scale is 8, so we know that such an output will take 8 hours. We can now move the cursor to 35 on the B scale, although this number is not in fact shown. Above it on the A scale is the figure 7, so an output of 350 tons will take 7 hours. Because all the

60 Calculating

settings are in the same proportion we can find the time needed for any output simply by moving the cursor.

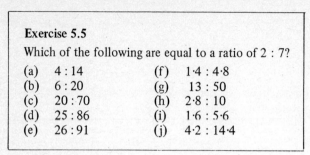

> **Exercise 5.5**
> Which of the following are equal to a ratio of 2 : 7?
> (a) 4 : 14   (f) 1·4 : 4·8
> (b) 6 : 20   (g) 13 : 50
> (c) 20 : 70   (h) 2·8 : 10
> (d) 25 : 86   (i) 1·6 : 5·6
> (e) 26 : 91   (j) 4·2 : 14·4

Interpreting graphs

Graphs are a method of presenting information without words. We must learn how to read them.

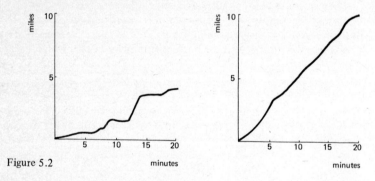

Figure 5.2

Figure 5.2 shows two graphs describing the journeys made by two men going to work by car. One is a town dweller and the other lives in the country. You will easily see which is which, but how much more information can you detect?

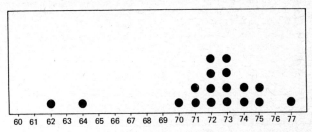

Figure 5.3

Figure 5.3 shows a distribution. Each dot represents a contestant at the start of a well-known springtime sporting event. What is their average age and what colours are they wearing?

Chapter 6

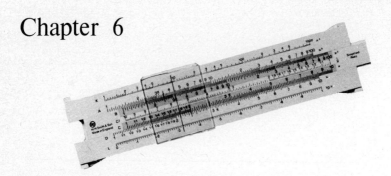

Slide rules are sometimes described as 'guessing sticks'. This is a useful description because they provide adequate answers to many problems but are less accurate than calculators and computers.

Powers of 2

In Chapter 4 we saw that we could use our table of powers of 2 to find square roots, cube roots, fourth roots, fifth roots, and so on. We found, for instance, that since $256 = 2^8$, then $\sqrt{256}$ was $2^{8 \div 2}$, that is, 2^4, or 16.

In Chapter 5 we saw that $\sqrt{2}$, although it cannot be calculated exactly, is approximately 1·4. We should be able to fit this number into our table of powers of 2. Set out as in Chapter 4 (page 34) we would have

2	2^1
$\sqrt{2}$	$2^{1 \div 2}$
1·4 approx.	$2^{\frac{1}{2}}$

We began this book by folding a piece of paper. Then we wrote the number of folds as indices: for example, 2^5, 2^{15} 2^{20}. Now we have a new index, $2^{\frac{1}{2}}$, which does not make sense when applied to folding paper, since we cannot fold a piece of paper half a time.

If, however, we take our simple slide rule marked with powers of 2, we can find a position for $2^{\frac{1}{2}}$ because the spaces correspond to the indices (one space gives 2^1, three spaces 2^3, and so on). $2^{\frac{1}{2}}$ must be halfway between 1 and 2. This can now be marked in, as in Figure 6.1.

Figure 6.1 1 ⊢ 1·4 2 ──── 4 ──────── 8 ──────── 16 ──────── 32

We can now use the slide rule procedure for multiplying, and we will find that $1·4 \times 1·4 = 2$. The halfway marks in the other spaces can be calculated very simply because when we add one space to another on the slide rule we are multiplying the numbers. If we add one space to half a space we get $2 \times 1·4$, that is $2·8$. We can multiply $4 \times 1·4$ ($2^2 \times 2^{\frac{1}{2}}$) by adding 2 spaces and half a space.

We must not forget that $1·4$ is an approximation and that we will get larger errors as we multiply by larger numbers. However, this example will help us to understand how the originators of the slide rule calculated the distances between the markings.

Fractions: division

You no doubt learnt at school that in order to divide by a fraction, you turn it upside down and multiply. This is true, but do you understand why? Look at Figure 6.2.

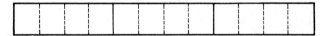

Figure 6.2

This represents 3 metres of wood marked off to make shelves each $\frac{1}{4}$ metre long. We can see at a glance that this will make 12 shelves.

If we start again with this problem and set it out mathematically we must ask how many $\frac{1}{4}$ metres there are in 3 metres. In sign language this problem becomes $3 \div \frac{1}{4}$ or $\frac{3}{1} \div \frac{1}{4}$. We know that 4 shelves can be cut from each metre and therefore 3×4 can be cut from 3 metres. So we know that we must finish the fraction calculation by writing $\frac{3}{1} \times \frac{4}{1}$. The answer will be $\frac{12}{1}$ or 12 shelves.

Now supposing we want shelves $\frac{3}{4}$ metre long. Obviously we will not get so many from the same length of wood. We know that we have 12 quarter metres, so we must divide by 3. To divide we put the number at the bottom of the fraction. So the problem of finding how many $\frac{3}{4}$ metres can be cut from 3 metres is written

$$\tfrac{3}{1} \div \tfrac{3}{4} = \tfrac{3}{1} \times \tfrac{4}{3} = \tfrac{12}{3} = 4$$

Mixed numbers are whole numbers and fractions combined, for example, $2\frac{1}{2}$. When multiplying and dividing with these numbers we change them by cutting up the whole ones to match the fraction. The number 2 in our example becomes 4 halves and is added to the $\frac{1}{2}$, that is, $2\frac{1}{2} = \frac{5}{2}$.

64   Calculating

In this form multiplication and division is simple, as all numbers are either above or below the line, as in the following

$$3\tfrac{2}{5} \div 2\tfrac{3}{7}$$
$$= \tfrac{17}{5} \div \tfrac{17}{7}$$
$$= \tfrac{17}{5} \times \tfrac{7}{17}$$
$$= \text{(by cancelling 17)} \ \tfrac{7}{5}$$
$$= 1\tfrac{2}{5}$$

Notice that the answer is changed back to a mixed number if possible.

Exercise 6.1

Think of these problems in the following way, for example: 'How many pieces $\tfrac{1}{6}$ (of a foot) in length can I cut from $\tfrac{2}{3}$ (of a foot) in length?'

Find:

(a) $\tfrac{2}{3} \div \tfrac{1}{6}$ (f) $2\tfrac{1}{2} \div \tfrac{1}{2}$
(b) $\tfrac{7}{8} \div \tfrac{1}{4}$ (g) $3\tfrac{2}{3} \div \tfrac{1}{3}$
(c) $\tfrac{8}{9} \div \tfrac{2}{3}$ (h) $3\tfrac{1}{5} \div \tfrac{4}{5}$
(d) $\tfrac{3}{4} \div \tfrac{1}{16}$ (i) $10 \div 1\tfrac{1}{2}$
(e) $\tfrac{5}{12} \div \tfrac{15}{16}$ (j) $6\tfrac{1}{4} \div 4\tfrac{2}{7}$

The examples in Exercise 6.1 have been chosen so that they can conveniently be 'cancelled out'. In real life problems involving fractions seldom cancel out so neatly. The advantage of using common fractions is therefore lost.

On a slide rule or a calculator we use decimal fractions. Thus 3 divided by $\tfrac{1}{4}$ becomes $\tfrac{3}{\cdot 25}$ and 3 divided by $\tfrac{3}{4}$ becomes $\tfrac{3}{\cdot 75}$. We are changing over to the metric system precisely to avoid the difficulties involved in dealing with common fractions.

Graphs

Data is a Latin word meaning 'given' and is used, for example, in the expression *experimental data*, meaning facts which are found by experiment. Experimental data are often arranged in the form of a graph which will show whether there is a relationship between the facts and what that relationship is.

Suppose we have a factory producing a constant flow of units of output and that the production rate may vary between 1000 and 1100 units per week. Records are kept for 10 weeks showing the weekly output in units, the number of accidents to work people and the number of units rejected as below specification.

Figure 6.3 shows the number of accidents plotted against the rates of weekly output. From this graph it does not seem that higher rates of output result in an increased number of personal accidents. If there were a direct relationship between the rate of output and the number of personal accidents, the crosses on the graph would form a clear pattern.

The comparison between the rate of output and the number of units rejected is shown in Figure 6.4. In this case the dots make a recognizable curve, which suggests that higher output necessarily means an increase in the number of rejected units.

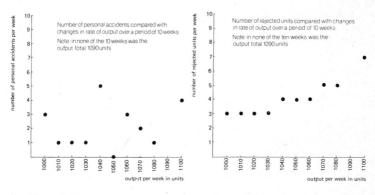

Figure 6.3

Figure 6.4

The slide rule

We have seen how to use the slide rule to multiply, but the examples in Exercise 4.6 were chosen because they gave exact answers. This is not always the case. Multiply on the slide rule 2·5 by 21. You will find that the answer lies between 52 and 53. The best you can say is that the answer is probably halfway between 52 and 53. In this case it is permissible to give the answer as 52·5. In all other cases the answer will be given correct to the nearest mark on the slide rule. Do not give an answer which appears to be more accurate than is in practice possible.

66 Calculating

When dividing on the slide rule remember that we subtract distances instead of adding them. Suppose we take 99 ÷ 3·3. The first thing we must do is to estimate, so we take 100 ÷ 3, which is approximately 30. Move the cursor to place the hairline over 99 on the A scale. Now move the slide so that 3·3 on the B scale lies under the hairline. Move the cursor back along the B scale to mark 1. You will see that above it is 30 on the A scale. So 99 ÷ 3·3 = 30, the same as our estimate.

While the slide rule is set in this position notice that, although the procedure is different, the same setting results from multiplying 30 × 3·3. See Figure 6.5.

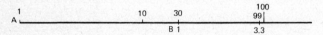

Figure 6.5

In fact you have established that the ratios $\frac{30}{1}$ and $\frac{99}{3\cdot3}$ are the same. The multiplication and the division are simply two different ways of looking at the same relationship.

Exercise 6.2

Use the A and B scales of the slide rule to calculate the following.

Always estimate first. Give the answer correct to the nearest mark on the slide rule.

(a) 4·1 × 14
(b) 24 × 2·9
(c) 17 × 4·5
(d) 3·6 × 9·2
(e) 5·3 × 18
(f) 92 ÷ 11
(g) 86 ÷ 6·1
(h) 75 ÷ 22
(i) 68 ÷ 6·8
(j) 59 ÷ 1·7

Calculators

There are many different models of small, electronic calculators. Here is an example of a simple keyboard. There is also an illuminated display for the results of calculations. The numbers and the signs −, +, ÷, ×, =, speak for themselves. The key on the bottom row with · is the decimal point. The key marked 'C' on the top row clears all entries whilst the key marked 'CE' clears only the last entry.

Detailed instructions are always supplied with calculators and, because each design differs slightly, it would be misleading to attempt to describe their operation here. However, we can show their essential simplicity by giving an example on the assumption that the keyboard is as illustrated.

The important thing to remember is that the keys are depressed in the order that you would state the problem in words.

For example, suppose we want to find 20 250 multiplied by 0·105. The process will be as follows.

1. Depress the keys for 2, 0, 2, 5, 0 in that order. Note that the figures and the decimal point in the window move to the left as each key is depressed.
2. Depress the key for multiply (×).
3. Depress the decimal point key followed by 1, 0, 5 in that order.
4. Depress the key marked + =.

The answer will appear in the illuminated display window.

Chapter 7

Some slide rules have the graduations on the circumference of a circle. Because the circumference is approximately 3·14 times the diameter there is more space for graduations and the circular slide rule is easier to read.

More powers

Your paper slide rule is restricted in its use by the fact that it is limited to powers of 2.

Make a series of powers of 3. Remember to begin with 1: $3^0 = 1$; $3^1 = 9$, and so on. You need not go beyond 1000.

Do the same for powers of 5, 7 and 10.

It would be convenient if we could put these on the same slide rule as our powers of 2. The problem would be to choose spaces for each of these series so that all the numbers fall into their correct places in the lines. You might like to try, but you will find it difficult to be accurate enough to be useful.

The scale in Figure 7.1 provides the line you will need.

```
1              2         4              8
|——————————————|—————————|——————————————|
```
Figure 7.1

First draw a line twice as long as this scale and copy these marks at the left-hand end. The first powers of 2 (from 2^0 to 2^3) are shown. By marking spaces equal to those separating these powers of 2 along your line you can put in powers of 2 up to 2^6.

Now put in the figures 3 and 9. You will find that the distance from 1 to 3 is equal to the distance from 3 to 9. Mark this distance twice more, which will give you positions for 3^3 (27) and 3^4 (81), and then you will have accurately fitted the powers of 3 into the scale of the powers of 2.

Following the same procedure, you will be able to put in the powers of 5. By marking off the distance from 1 to 5 again, you will find the position of 5^2 (25).

Similarly you can find 6^2, 7^2 and 10^2 (4^2, 8^2 and 9^2 you will find already marked).

In Chapter 5 (page 58) we calculated the square root of 2 ($2^{\frac{1}{2}}$). In Chapter 6 (page 62) we marked it halfway between 1 and 2 on our scale. It will be useful also to find the square root of 10 ($10^{\frac{1}{2}}$). In Figure 7.1 the distance from 1 to 10 (10 is not marked) is 10 centimetres (cm) or 1 decimetre (dm). We can find the midpoint between 1 and 10. It will be 5 cm or 0·5 dm. This is $10^{\frac{1}{2}}$ or $\sqrt{10}$. We will find it comes just after 3 so we can guess that $\sqrt{10} \approx 3\cdot2$. We now write $10^{\frac{1}{2}}$ as $10^{0\cdot5}$.

Now if 3·2 can be written as $10^{0\cdot5}$ we can also find the numbers 2, 3, 4, etc., as powers of 10. If we measure the distance from 1 to 2 we find it is 3. This is $\frac{3}{10}$ dm, or 0·3 dm. So $2 = 10^{0\cdot3}$. From 1 to 3 is 4·8 cm (48 mm), that is, $3 = 10^{0\cdot48}$.

We can now find all the other numbers as powers of 10 and set them down in a table.

Number	Powers of 10
1	10^0
2	$10^{0\cdot3}$
3	$10^{0\cdot48}$

Complete the numbers from 1 to 10 for yourself. You should finish with $10 = 10^1$.

We can now multiply using powers of 10, as we did in Chapter 1, using powers of 2. Here is a simple example

Number		Power of 10
2	=	$10^{0\cdot3}$
× 3	=	$10^{0\cdot48}$
	=	$10^{0\cdot78}$

From your table you can find that $10^{0\cdot78}$ is 6, which is obviously the correct answer.

Decimals: four rules

You may sometimes have to work without the aid of slide rules and calculators. Here are the rules.

70   Calculating

Addition and subtraction

Simply put the figures under one another in their correct place on the number line. Here are two examples:

$$27·4 + 623 + 2·04 \qquad 369 - 14·321$$

```
    27·4            369
   623            14·321
    2·04          ──────
  ──────         354·679
  652·44
```

Exercise 7.1

Addition:

(a) 42·1 + 4·21
(b) 80·5 + 127
(c) 1006 + 250·8
(d) 62 + 4·01 + 127
(e) 26·403 + 500·004 + 32·6

Exercise 7.2

Subtraction:

(a) 14·297 − 13·937
(b) 132·04 − 21·835
(c) 6·928 − 5·604
(d) 95·3 − 4·938
(e) 127 − 0·999

Multiplication

Make an estimate first.

Put the multiplier under the top figure so that the whole numbers are in line with the right-hand end of the top line as in the examples below. All the other figures will then appear in their correct places on the number line.

6·201 × 32

Estimate: 6 × 30 = 180

```
              6·201
           ×     32
          ─────────
   × 30   186·030
   ×  2    12·402
   ────   ────────
     32   198·432
```

6·201 × 32·5

Estimate: 6 × 30 = 180

```
               6·201
           ×    32·5
          ──────────
   × 30    186·030
   ×  2     12·402
   × 0·5     3·1005
   ─────   ─────────
     32·5  201·5325
```

Chapter 7 71

Exercise 7.3

Multiplication:
(a) 42·7 × 0·34
(b) 8·51 × 8·21
(c) 39·04 × 1·02
(d) 625 × 5·13
(e) 21·04 × 925

Division

Make an estimate first.

Multiply the top and bottom numbers by 10, 100 or 1000, etc., to get a whole number in the denominator. This will help you to check your estimate.

The figures in the answer will be in their correct place on the number line. As in short division, 0 can be added after the decimal point to give the required number of places of decimals.

Example: $\dfrac{35·43}{3·5}$

Estimate: $30 \div 3 = 10$

$$\dfrac{35·43 \times 10}{3·5 \times 10} = \dfrac{354·3}{35}$$

```
           10·12
      35 ) 354·30
           35
           ‾‾
           04 3
            3 5
           ‾‾‾
              80
              70
              ‾‾
              10
```

Answer: 10·12, or 10·1 correct to one place of decimals

Example: $\dfrac{354·3}{0·35}$

72 Calculating

Estimate: $300 \div \dfrac{3}{10}$

$= \dfrac{300 \times 10}{3}$

$= 1000$

$\dfrac{354 \cdot 3 \times 100}{0 \cdot 35 \times 100} = \dfrac{35\,430}{35}$

```
            1012·28
       35 ) 35430·00
            35
            ___
            043
             35
             __
             80
             70
             __
             30 0
             28 0
             ____
              2 0
```

Answer: 1012·28, or 1012·3 correct to one place of decimals

Exercise 7.4
Division:
(a) $31 \cdot 68 \div 1 \cdot 2$
(b) $43 \cdot 369 \div 5 \cdot 4$
(c) $759 \cdot 6 \div 3 \cdot 6$
(d) $70 \cdot 455 \div 7 \cdot 7$
(e) $286 \cdot 3 \div 3 \cdot 54$

The slide rule

Look at the C and D scales of the slide rule. Between 1 and 2 we have numbers 1·1, 1·2, 1·3, etc., and these are subdivided so that we can find 1·03, 1·18, 1·27, and so on. It is for this reason more accurate than the scales A and B. The space between 2 and 3 is also subdivided so that we can read not only 2·4, 2·9, etc., but 2·02, 2·04,

Chapter 7   73

2·06, 2·08, and so on. In fact, along the whole scale there are twice as many markings as between the same figures on the A and B scales.

The C and D scales only go from 1 to 10 and this would seem to limit what you can do, but, as we saw in Chapter 6, we can read these markings as numbers from 1 to 10, 10 to 100, or 100 to 1000, and so on as we wish.

The importance of first making an estimate will soon be recognized when using the C and D scales to multiply numbers larger than 10.

Take this example

$$232 \times 10·9 \quad \text{Estimate: } 200 \times 10 = 2000$$

When setting the slide rule always ignore the decimal point in your figures. In this case read the figures as though 1 stands for 100 and 10 stands for 1000. With the cursor mark 232 on the D scale. Move the slide so that 1 on the C scale is under the hairline. Re-set the cursor so that the hairline marks 109 on the C scale. The answer is under the hairline on the D scale, halfway between 252 and 254, so call it 253. Now look back to our estimate. It was 2000. The answer must therefore be 2530 correct to three significant figures. The correct answer is 2528·8, so we are only out by 1·2.

Here is another example

$$4·05 \times 18·6 \quad \text{Estimate: } 4 \times 20 = 80$$

Using the slide rule as before we find the nearest mark is 75. Our original estimate was 80, so we can give the answer as 75 correct to two significant figures. The exact answer is 75·33. Again the slide rule error is very small.

The estimates in these two examples were not very close to the correct figure, but they did serve to warn us that the answer in one case could not be 253 or 25·3 but was 2530, and in the other, not 750 or 7·5 but 75.

One more difficulty remains when using the C and D scales, for example:

$$38 \times 41 \quad \text{Estimate: } 40 \times 40 = 1600$$

Using the cursor put 1 on the C scale above 38 on the D scale. If you move the cursor to 41 on the C scale you will find it runs off the end of the slide rule. Move the slide back so that 10 on the C scale is above 38 on the D scale. Now re-set the cursor to 41 on the C scale. The hairline marks the answer 156 on the D scale. The estimate was

1600 so the slide rule answer must be 1560 correct to three significant figures.

By putting 10 instead of 1 above 38 on the D scale we are using the D scale as though it were the right-hand end of a double scale like the B scale. With a circular slide rule the re-setting is not necessary, because you can repeat the scale by going round again without interruption.

Mark a strip of paper with numbers 1 to 10 spaced as on scale D and also mark 38. Add this to the left-hand end of your D scale. You will see that 1 on the C scale is now over 3·8.

Exercise 7.5

Calculate the following using the C and D scales of your slide rule and express your answers as correct to two or three significant figures:

(a) 1·5 × 23
(b) 2·8 × 1·7
(c) 32 × 54
(d) 65 × 1·2
(e) 5·4 × 6·1
(f) 855 × 1·14
(g) 23·2 × 72·9
(h) 63·5 × 4·3
(i) 74·5 × 0·915
(j) A railway signalman has a basic rate of pay of £27·60 per week. What would be his total basic pay (before deductions) for a year?

Chapter 8

The slide rule finds the answer to a multiplication by adding spaces together. Logarithm tables give numerical values for the spaces.

Logarithms

In the first section of Chapter 7 we found that the numbers from 1 to 10 could each be written as a power of 10. We found that the index numbers were 0, 0·3, 0·48, and so on.

These numbers are also called 'logarithms' ('logs' for short). Instead of using the expression '$2 = 10^{0·3}$', we could say 'the logarithm to the base 10 of the number 2 is 0·3'. Logarithm tables are based on 10, and we can multiply or divide by adding or subtracting the logarithms of the numbers. The scale in Figure 7.1 is known as a logarithmic scale.

We frequently wish to illustrate, by means of a graph, a succession of numerical changes, say in the cost of living index or rates of wages. If, for this purpose, we use ordinary squared paper we shall find that the line showing, for example, the change from 100 to 110 will have the same slope as the change from 150 to 160. This is misleading because in the first case the percentage change is 10%, whereas between 150 and 160 the percentage change is about $6\frac{1}{2}\%$.

Figure 8.1 is a graph of the changes in wage rates over a period of

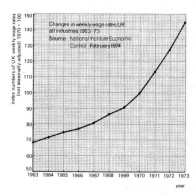

Figure 8.1

75

years on ordinary squared paper. In Figure 8.2 the same figures are plotted on paper on which the vertical axis is graduated on a logarithmic scale and correctly illustrates the rate of change in wage rates over the period.

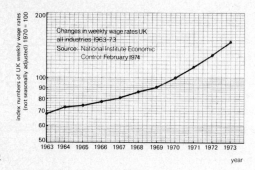

Figure 8.2

Ratio and proportion

In Chapter 1 we showed two wheels in the ratio of 9 : 3. We saw that this could be written $\frac{9}{3}$. We could have drawn the wheels in the opposite order, with the smaller on top. In this case the ratio would have been written 3 : 9 or $\frac{3}{9}$.

George and young David produce components in this ratio

$$\frac{\text{David}}{\text{George}} = \frac{7}{8}$$

If David's piece-rate bonus is £6·30 and the bonuses are directly proportional to output we can find George's bonus. It is convenient in this case to put George's bonus first

$$\frac{\text{George's bonus}}{£6·30} = \frac{8}{7}$$

$$\text{George's bonus} = \frac{8 \times £6·30}{7} = £7·20$$

Any set of four numbers that combine to make two equal ratios can, in fact, be written in four different ways. For example

$$\frac{1}{2} = \frac{8}{16} \qquad \frac{2}{1} = \frac{16}{8}$$
$$\frac{8}{16} = \frac{1}{2} \qquad \frac{16}{8} = \frac{2}{1}$$

Suppose we want to find the cost of 3·6 metres of wood when

1 metre costs 18p. Putting x pence for the cost of 3·6 metres we can write the ratio in four different ways

$$\frac{3·6\,\text{m}}{1\,\text{m}} = \frac{x\,\text{p}}{18\,\text{p}} \quad \text{or} \quad \frac{1\,\text{m}}{3·6} = \frac{18\,\text{p}}{x\,\text{p}} \quad \text{or} \quad \frac{x\,\text{p}}{18\,\text{p}} = \frac{3·6\,\text{m}}{1\,\text{m}} \quad \text{or} \quad \frac{18\,\text{p}}{x\,\text{p}} = \frac{1\,\text{m}}{3·6\,\text{m}}$$

The most convenient way for the purposes of calculating is the one that begins with the unknown quantity

$$\frac{x\,\text{p}}{18\,\text{p}} = \frac{3·6\,\text{m}}{1\,\text{m}}$$

$$x = \frac{3·6 \times 18}{1}$$

$$= 64·8 \ (65)$$

The cost of 3·6 metres of wood is 65p.

Exercise 8.1

Find the missing terms:

(a) $\dfrac{4}{5} = \dfrac{20}{x}$

(b) $\dfrac{8\,\text{men}}{x\,\text{men}} = \dfrac{£120}{£150}$

(c) $\dfrac{3\,\text{lb}}{8\,\text{lb}} = \dfrac{13\frac{1}{2}\,\text{p}}{x\,\text{p}}$

(d) $\dfrac{£x}{£40} = \dfrac{105}{100}$

(e) The special bonus rate is £3·30 per hour. If a man earns a total of £16·50, how many hours has he worked?

Percentages: up and down

Increase £100 by 10% and then decrease the result by 10%. How much have you now? The answer is not £100.

10% of £100 is £10. After increasing £100 by 10% the result is £110. 10% of £110 is £11. After decreasing £110 by 10% the result is £99.

78 Calculating

> **Exercise 8.2**
>
> (a) On Monday the hens laid 12 eggs. On Tuesday there was a 50% increase. On Wednesday there was a decrease of 50% from Tuesday's figure. How many eggs were laid on Wednesday?
>
> (b) After months of dieting Bill (16 stone) decreased his weight by $12\frac{1}{2}\%$. He soon increased again by $12\frac{1}{2}\%$. How much did he weigh then?
>
> (c) Production for January was 1000 units. The February figure was 20% up on January's. In March there was a decrease of 20% on February's figure. How many units were produced in March?

Simple interest

Interest is a kind of rent which a borrower pays for the use of the money lent to him. This is known as *simple interest*. One sort of interest is the money you receive if you lend money to the government. Suppose you invest £100 in government stock (often called 'gilt-edged'). The government will undertake to pay you the interest on this amount at the rate of 8% per annum ('*per annum*' is Latin for 'each year'); the rate changes from time to time. You will be paid £8 every year, although normally in fact you will be paid £4 each half year.

Calculating simple interest

If the amount invested is in round hundreds this is easy. For instance £500 at 8% will give £8 × 5 or £40; £6000 at 10% will give £10 × 60 or £600.

If the capital is not in round hundreds you can still find out how many hundreds there are by using a decimal point. For instance, if the capital amount is £650 and the interest 8%, you would put the decimal point after the 6 because you have 6 hundreds, and so calculate 8 × 6·50. You have in fact $6\frac{1}{2}$ hundred pounds and therefore you will multiply the rate of interest by $6\frac{1}{2}$.

The use of decimal fractions makes it easy to calculate complicated amounts, especially if you have a calculating machine. For instance

$$\text{Interest on £2450 at 6\%} = £6 \times 24·50$$

Interest on £322 at 4% = £4 × 3·22
Interest on £10 068 at 9% = £9 × 100·68

If the amount invested includes pence we must move the decimal point so as to divide by 100. For example

£200·50 at 5% would be £5 × 2·005
£520·75 at 8% would be £8 × 5·2075
£362·25 at 12% would be £12 × 3·6225

The result of £12 × 3·6225 is £43·47.

Second method

When discussing percentages we showed that 1% of £1 is 1p and 8% of £1 is 8p and so on. We can therefore calculate simple interest this way. For instance

£1 at 5% would be 5p or £0·05
£2 at 5% would be £0·05 × 2
£53 at 5% would be £0·05 × 53
£460 at 5% would be £0·05 × 460

It will be a good idea now to find the interest on £1 for one year. In a formula this is called 'i'.

Complete the following table.

Rate of interest	i
5%	0·05
$2\frac{1}{2}$%	0·025
20%	0·20
$3\frac{1}{2}$%	0·035
$4\frac{1}{4}$%	0·0425
10%	
$8\frac{1}{2}$%	
$12\frac{1}{2}$%	
$11\frac{1}{4}$%	
$9\frac{3}{4}$%	

Our formula is now simple. Using the letter 'P' for the principal (that is the amount of capital invested) and the letter 'I' for the amount of interest you expect to receive in one year,

$$I = P \times i$$

We can now calculate the amount of interest we would receive for any length of time by multiplying by the number of years, which we can call 't'.

The formula becomes $I = P \times i \times t$, or more usually $I = Pit$.

$I = Pit$ replaces the old formula $I = \dfrac{PRT}{100}$ where R stood for the rate of interest. By changing R, which would be say 5, to i, which would be 0·05, we have already divided by 100 and the calculation can easily be done on a machine by one operation.

Exercise 8.3

Find the simple interest on:
(a) £300 at 5% for 2 years
(b) £850 at 6% for 3 years
(c) £1284 at $12\frac{1}{2}$% for 7 years
(d) £2500 at $9\frac{1}{4}$% for 20 years
(e) £8650 at $11\frac{1}{2}$% for 25 years

Sometimes interest problems come the other way round. Suppose you were told that the interest for one year on £250 was £17·50 and you needed to know the rate of interest. Since this is for one year the formula is $I = Pi$. If we insert the figures we already know, we get

$$17 \cdot 50 = 250i$$

We can write this, as we did when calculating ratios, as

$$\frac{17 \cdot 50}{1} = \frac{250i}{1}$$

Change the order to $\dfrac{250i}{1} = \dfrac{17 \cdot 50}{1}$ to bring the number we are trying to find to the left-hand side. To get i by itself we can now divide both sides by 250.

$$i = \frac{17 \cdot 50}{250}$$

$$= 0 \cdot 07$$

The rate of interest is therefore 7%.

If we have a number of years to consider, the method is the same. For example, to find i when the principal is £174 and the interest for three years is £62·64

$$I = Pit$$
$$62·64 = 174 \times i \times 3 = 62·64$$
$$i = \frac{62·64}{174 \times 3}$$
$$= 0·12$$

Exercise 8.4

Find i (the amount of interest on £1 for one year):

(a) Principal £160, interest £4
(b) Principal £400, interest £16
(c) Principal £450, interest £20·25
(d) Principal £760, interest £68·40
(e) Principal 1500 francs, interest 97.5 francs

The slide rule

Division

You will remember that the sign ÷ really represents a fraction and that we often write $\frac{4}{2}$ rather than $4 \div 2$. This method of writing divisions is helpful when doing complicated calculations, in writing formulae and in solving ratio problems. It is particularly helpful with the slide rule. Set 4 above 2 on the C and D scales and you will find the answer over the 1. What is more you will find that all along the scale you have divisions that give the same answer.

$$\frac{6}{3} \quad \frac{8}{4} \quad \frac{10}{5}$$
$$\text{and even} \quad \frac{3}{1·5} \quad \frac{5}{2·5} \quad \text{and so on}$$

Now set 2 above 4. This time the 1 on the D scale is off the end of the C scale. However, there is no need this time to move the slide. Simply look for the figure above the 10. It is 5 in this case. We know that $\frac{2}{4}$ is less than 1, so the answer must be 0·5.

It is still very important to estimate the answer so as to get the decimal point in the correct place. Here are some examples

(a) $\frac{58}{41} \approx \frac{60}{40} \hat{=} 1\cdot\text{something}$

(b) $\frac{32}{53} \approx \frac{30}{50} = 0.6$

(c) $\frac{10}{8\cdot 42} \approx \frac{10}{8} \hat{=} 1\cdot\text{something}$

(d) $\frac{6\cdot 2}{670} \approx \frac{6}{600} = \frac{1}{100} = 0.01$

(e) $\frac{3\cdot 7}{9540} \approx \frac{4}{10\,000} = 0.0004$

In the last two examples we have fractions with denominators of 100 and 10 000 and we must move the decimal point the same number of places as the number of noughts in the denominator to give the decimal equivalent.

Exercise 8.5

Use the slide rule to give the decimal equivalents of the fractions (a)–(e) above.

Make estimates for the following and give the answers from the slide rule:

(f) $\dfrac{45}{2\cdot 82}$

(g) $\dfrac{3}{47\cdot 5}$

(h) $\dfrac{5\cdot 6}{180}$

(i) $\dfrac{130}{7\cdot 65}$

(j) $\dfrac{8\cdot 9}{3800}$

Ratios and the slide rule

Earlier we discussed the problem of finding the missing number in a problem set down as a proportion. We found that we had four ways of writing down the two ratios and chose the one that was most convenient. Now if we set our problem on a slide rule we find that it does not matter which arrangement we choose.

Our example was: 1 metre of wood cost 18p; find the cost of 3·6 metres. The ratio of lengths is $\frac{3\cdot 6}{1}$ and the cost is proportional, so we wrote $\frac{3\cdot 6}{1} = \frac{x}{18}$. By setting $\frac{3\cdot 6}{1}$ on the slide rule we can find the value of x over 18. It is approximately 65. If we take the *inverse* (that is, if we turn the ratio upside down) and set $\frac{1}{3\cdot 6}$ we would find x this time below 18.

Percentages on the slide rule

In certain cases a number of percentages can be found with only one setting of the slide rule.

Screws are produced in batches of seventy-two. In one batch there are nine rejects. That is a ratio of $\frac{1}{8}$ or $\frac{12 \cdot 5}{100}$. We could write this as follows

$$\frac{9}{72} = \frac{12 \cdot 5}{100}$$

This ratio can be rearranged as

$$\frac{9}{12 \cdot 5} = \frac{72}{100}$$

The slide rule is set with 72 above 100. The number of rejects, nine, can be read off as $\frac{12 \cdot 5}{100}$ or 12·5%. Without changing the setting of the slide rule the number of rejects in other batches of screws can be read off as percentages.

Exercise 8.6

Give the following as percentages, correct to two significant figures:

(a) $\frac{35}{65}$ (b) $\frac{58}{65}$ (c) $\frac{26}{65}$ (d) $\frac{52}{65}$ (e) $\frac{48}{65}$

Chapter 9

The thermometer gives a practical illustration of the use of negative numbers.

Negative numbers

Negative numbers are those below zero. They are used in recording temperatures. The freezing point of the Fahrenheit scale is 32° and the boiling point is 212°. We now use Centigrade with 0° for freezing point and 100° for boiling point. It is much simpler to have the amount of frost stated as simply $-2, -5$, and so on.

These negative numbers in mathematics cause confusion because we are using the minus sign, $-$, in two different ways. It may refer to doing something, that is, taking away, when it is a verb. It may also describe the kind of number we are talking about as 'negative', in which case it is an adjective.

Sometimes it is convenient to distinguish between these two ways of using the minus sign. Suppose in the game described in Chapter 3 we write down the scores thrown with the dice. We can write the blue numbers as $^+1, ^+2, ^+3$, etc., and red numbers as $^-1, ^-2, ^-3$, etc. We can then write the score by adding two together, for example

$$^+3 + {^+1} = {^+4}$$
$$^+3 + {^-1} = {^+2}$$

What if we want to subtract? This is simple

$$^+3 - {^+1} = {^+2}$$

Logically,

$$^-3 - {^-1} = {^-2}$$

If you have 3 red numbers and take away 1 red number you are left with 2 red numbers.

What would happen in this case?

$$^+3 - {}^-1$$

Put a counter on 3 ($^+3$) in a number line as in Figure 9.1.

Figure 9.1   -5   -4   -3   -2   -1   0   1   2   ③   4   5

If you now have to add a score of $^-1$ it would bring your score down to 2, that is

$$3 + {}^-1 = 2 \quad \text{(see Figure 9.2)}$$

Now suppose you have moved out of turn and want to remove the last score. You are at the moment on 2.

Figure 9.2   -5   -4   -3   -2   -1   0   1   ②   3   4   5

By removing your negative score you will come back to your original position on 3, so we now have

$$2 - {}^-1 = 3$$

The effect of subtracting (taking away) the negative score $^-1$ is to add 1.

When you write a cheque for £5 you take £5 off your balance. For you it is $^-$£5. If you then find that for some reason the cheque has not been cashed you remove this item by taking away the $^-$£5. You will be £5 better off than before.

$$- {}^-£5 = {}^+£5$$

Take a set of counters in two colours; small squares of paper will do, but buttons or plastic counters will be easier to handle. You will need about twenty of each. Suppose we have white counters for positive (+) numbers and black counters for negative (−) numbers. Put ten of each out on the table in front of you as in Figure 9.3. The total of +10 and −10 is 0.

Figure 9.3

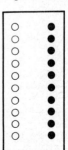

86   Calculating

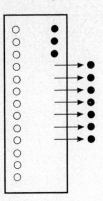

Example 1:

To find $3 - {}^-7$, first put out 3 more white counters, then take away 7 black ones (Figure 9.4). You now have 3 white counters paired off with 3 black ones and 10 extra white ones. This is a score of $+10$.

Figure 9.4

Example 2:

If we want to work $3 + {}^-7$, we must start with 10 of each; add 3 white and then add 7 black. We will get 4 black ones not paired off, an answer of $^-4$. You can now work Exercise 9.1 using these counters.

> **Exercise 9.1**
> A number that has no sign in front of it is always positive.
>   Find:
> (a) $8 + {}^-5$ (f) $9 - {}^+4$
> (b) $7 + {}^+1$ (g) $9 - {}^-4$
> (c) $6 + {}^-8$ (h) $4 - {}^+8$
> (d) $3 + {}^-1 + {}^+2$ (i) $1 - {}^-6 - {}^+2$
> (e) $7 + {}^-9 + {}^-4$ (j) $0 - {}^-1 - {}^-8$

The slide rule

A series of calculations can be worked in succession on the slide rule without writing down the answer at each stage.
   Consider
$$\frac{1 \cdot 26 \times 2 \cdot 5 \times 3 \cdot 1}{4 \cdot 7}$$

Estimate
$$\frac{1 \times 3 \times 3}{5} = \frac{9}{5} \approx 2$$

Using the C and D scales the steps are as follows:

1. 126 × 25 = 315 on the D scale.
2. Leave the cursor hairline over 315 on the D scale.
3. Move the slide to multiply 315 by 31. This is 975 on the D scale.
4. Move the cursor to 975 on the D scale.
5. Move the slide to bring 47 on the C scale under the hairline.
6. We are dividing by 47. Therefore the answer is on the D scale below 1 on the C scale, that is, 208.
7. From the estimate we know that the answer to the problem is about 2. Therefore the slide rule answer is 2·08.

If you care to work it out by long multiplication and division you will find that the answer is 2·0776 correct to four decimal places which is 2·08 correct to two decimal places.

Note that the division $\frac{975}{47}$ was set on the slide rule as $\frac{47}{975}$. The answer appears as $\frac{1}{208}$ so we have $\frac{47}{975} = \frac{1}{208}$. You will remember that ratios set in this form are still equal when turned upside down, so $\frac{99}{47} = \frac{208}{1}$.

Exercise 9.2

Find the slide rule answers to the following:

(a) $\dfrac{31 \times 22}{8}$

(b) $\dfrac{48 \cdot 5 \times 3 \cdot 85}{4 \times 3 \cdot 5}$

The denominator of (b) consists of two numbers separated by a × sign. **Divide** by each in turn. (Dividing by 4 and then by 3·5 is exactly the same as dividing by 4 × 3·5, that is, 14.)

(c) $\dfrac{20 \cdot 7 \times 3 \cdot 1 \times 5 \cdot 1}{\cdot 89}$

(d) $\dfrac{142 \cdot 6 \times 15 \cdot 8 \times 1 \cdot 6}{41 \cdot 2}$

(e) What will be the interest on £150 000 at 7½% for 52 days?

Ratio and proportion

'If 4 men take 12 hours to do a job, how long will 6 men take?' Clearly the men and hours are in some kind of proportion. If, however, we say that if 4 men take 12 hours, then 6 men will take 18 hours, and write it like this,

>4 men   12 hours
>6 men   18 hours

the gang of 4 men might make some unprintable comments. In fact, of course, the more men there are the less time they will take — other things being equal.

Therefore we should write it this way

>4 men   12 hours
>6 men   ? hours

To make the ratios equal the time taken by 6 men is 8 hours. Because we have to turn one side upside down this is known as *inverse proportion*.

The simplest way to solve this problem is to use an equation. The total man/hours are the same in both cases. If h is the number of hours 6 men must work, we get

$$6 \times h = 4 \times 12$$

so $$h = \frac{4 \times 12}{6}$$

$$= 8$$

Exercise 9.3

(a) At 50 mph a journey takes 7 hours. How many hours will the same journey take at 7 mph?

(b) You have enough money to buy 7 gallons at 45p per gallon. How many gallons could you buy at 35p per gallon?

(c) A ship has enough stores to feed 28 men for 8 days. If 4 more men are picked up, how long will the food last?

(d) A wall can be covered by 120 6-inch square tiles (36 sq in). How many 4-inch square tiles would be needed for the same wall?

(e) A drive wheel at each revolution turns a cog wheel with 16 teeth $1\frac{1}{2}$ times. How many times would each revolution of the drive wheel turn a cog wheel with 12 teeth?

Graphs

So far the graphs we have used have been concerned with positive numbers only. There are many practical uses for graphs where negative numbers will not be required. Wages, however many deductions for insurance or tax, never go below zero. Graphs showing numbers of people cannot have negative numbers.

On the other hand, temperature can go below zero and the balance of payments can get 'into the red', so we need to have graphs which can show this.

First we need a horizontal number line with the 0 in the middle, as in Figure 9.5.

-10 -9 -8 -7 -6 -5 -4 -3 -2 -1 0 1 2 3 4 5 6 7 8 9 10

Figure 9.5

We also need a vertical number line as in Figure 9.6. Take a piece of graph paper and draw these two lines so that they cross in the middle of the page, and number this crossing 0 (see Figure 9.7).

Now put in the other numbers as in Figures 9.5 and 9.6 as far as you can.

The horizontal line we name x and the vertical y. The value of the y numbers will depend on the equation connecting x and y.

10
9
8
7
6
5
4
3
2
1
0
-1
-2
-3
-4
-5
-6
-7
-8
-9
-10

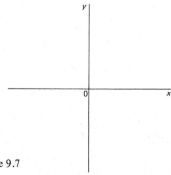

Figure 9.6

Figure 9.7

The simplest possible example shows the 2 times table. That is, we are going to make y equal to $2x$. We can list a few possible pairs

90   Calculating

of numbers, for example, if $x = 3$ then $y = 6$. Then we make a table.

x	y
3	6
4	8
−1	−2
−5	−10

When we place these numbers on the graph they will appear as in Figure 9.8 and, as you will see, a straight line can be drawn which passes through all these points.

You can add other figures to the table and then mark them on your graph. You will find that they all lie on the same line.

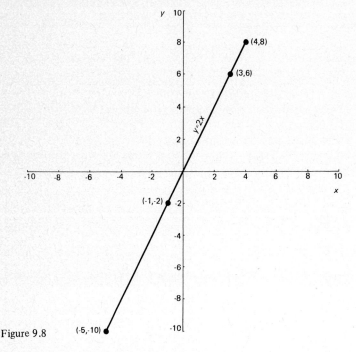

Figure 9.8

We can use this kind of graph to calculate other multiples of 2. For example, if we start where x is -3.5 we can find 2×-3.5 by drawing a dotted line to the graph line as in Figure 9.8. We can see that we reach the position $(-3.5, -7)$. That is, when the x number is -3.5 the y number is -7, so $2 \times (-3.5) = -7$. Find on your graph the values of y when x is 1.4, 2.6, -2.2 and -4.8.

Take a sheet of graph paper and draw a line for each of the equations in Exercise 9.4. (The equation to the line in Figure 9.8 is $y = 2x$.) Find three or four pairs for each equation as in the table on page 92. The notes at the side of each problem will help you.

Exercise 9.4
(a) $y = 10x$ [If $x = 0.5$, $y = 10 \times 0.5 = 5$]
(b) $y = 2x + 3$ [If $x = 0$, $y = (2 \times 0) + 3 = 3$]
(c) $y = 3x - 4$ [If $x = 1$, $y = (3 \times 1) - 4 = -1$]
(d) $y = 6 - 2x$ [If $x = 4$, $y = 6 - (2 \times 4) = -2$]

Notice the angles and the directions in which the lines slope. Could you write the equation to another line which would be parallel to $y = 2x + 3$? Could you also write the equation to a line sloping in the opposite direction? Add these lines to those done in Exercise 9.4.

Postscript

Take a sheet of paper. Put a dot halfway down and about $1\frac{1}{2}$ in from the left-hand edge. Make creases by folding the paper so that different points on the left-hand edge touch the dot (see Figure 9.9). When you have made a large number of creases you will find that they enclose (envelope) a curve. This is known as a parabola.

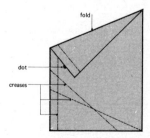

Figure 9.9

This is the shape of a car headlamp. Can you see how light would be reflected from a bulb in the position of the dot?

92 Calculating

The Jodrell Bank 'dish' is a similar shape in order to focus incoming radio waves on to a receiver.

Answers

Exercise 1.1
(a) 8 192
(b) 1 048 576
(c) 16 384
(d) 32
(e) 524 288

Exercise 1.2
(a) ⚶ ⚶
(b) ⚶ ⚶ ⚶
(c) ⚶ ⚶ ⚶
(d) ⚶ ⚶
(e) ⚶ ⚶ ⚶ ⚶

Exercise 1.3
(a) $\frac{7}{8}$
(b) $\frac{7}{16}$
(c) $\frac{17}{32}$
(d) $\frac{15}{32}$
(e) $\frac{23}{32}$

Exercise 1.4
(a) 7·3 in
(b) 12·8 in
(c) 0·4 in
(d) 32·6 in
(e) 15·5 in
(f) £5·43
(g) £2·06
(h) £0·01
(i) £8·75
(j) £8·75
(k) 6·894 m
(l) 3·204 m
(m) 0·007 m

Exercise 1.4 (contd)
(n) 0·532 m
(o) 0·068 m

Exercise 1.5
(a) 25%
(b) 20%
(c) 34%
(d) 80%
(e) 65%
(f) 30%
(g) 20%
(h) 5%
(i) 10%
(j) 10%

Exercise 2.1
(a) 11
(b) 1 010
(c) 1 111
(d) 1
(e) 1 100

Exercise 2.2
(a) 1 011
(b) 1 101
(c) 100
(d) 1 110
(e) 111

Exercise 2.3
(a) 4
(b) 2
(c) 8
(d) 1024
(e) 4096

Exercise 2.4
(a) $1\frac{7}{12}$
(b) $\frac{1}{12}$
(c) $\frac{5}{12}$
(d) $1\frac{1}{12}$
(e) $1\frac{7}{12}$
(f) 2

Exercise 2.5
(a) $\frac{4}{15}$
(b) $\frac{9}{10}$
(c) $\frac{19}{60}$
(d) $1\frac{11}{12}$
(e) $1\frac{19}{60}$
(f) $1\frac{9}{20}$

Exercise 3.1
(a) 262 144
(b) 1 048 576
(c) 65 536
(d) 262 144
(e) 262 144

Exercise 3.2
(a) 2048
(b) 16 384
(c) 524 288
(d) 1 048 576
(e) 1 048 576

Exercise 3.3
(a) 1·5
(b) 1·9
(c) 2·5

94 Calculating

Exercise 3.3 (contd)
(d) 7·6
(e) 9·1
(f) 5·3

Exercise 3.4
(a) 0·143
(b) 0·429
(c) 0·571
(d) 0·714
(e) 0·857

Exercise 3.5
(a) 700
(b) 30 000
(c) 350
(d) 21 000
(e) 210
(f) 37
(g) 582·1
(h) 642·4

Exercise 3.6
(a) 34
(b) 81·2
(c) 423
(d) 60·41
(e) 2018
(f) 3009

Exercise 3.7
(a) 1·5
(b) 2·74
(c) 8·1
(d) 12·8
(e) 102·4
(f) £0·01
(g) £0·05
(h) £0·37
(i) 1·37
(j) 1·208

Exercise 3.8
(a) 2
(b) 8
(c) 1 100 (12)

Exercise 3.8 (contd)
(d) 1 110 (14)
(e) 1 010 (10)
(f) 10 (2)
(g) 1 011 (11)
(h) 11 (3)
(i) 10 (2)
(j) 111 (7)

Exercise 3.9
(a) £13·72
(b) £69·65
(c) £72·57
(d) £38·79
(e) 150 tons

Exercise 3.10
(a) 5200 tons
(b) £32·40
(c) £55 300 million

Exercise 4.1
(a) 32
(b) 8
(c) 8
(d) 4
(e) 4
(f) 16
(g) 256
(h) 8
(i) 4
(j) 2

Exercise 4.2
(a) 42
(b) 63
(c) 33
(d) 118
(e) 77
(f) 83

Exercise 4.3
(a) 101 011
(b) 011 101
(c) 010 001
(d) 111 111

Exercise 4.3 (contd)
(e) 100 111
(f) 001 011
(g) 110 010
(h) 001 111
(i) 011 111
(j) 010 111

Exercise 4.4
(a) $\frac{3}{10}$
(b) $\frac{5}{14}$
(c) $\frac{1}{6}$
(d) $\frac{1}{15}$
(e) $\frac{1}{4}$
(f) $\frac{1}{9}$

Exercise 4.5
(a) 0·54
(b) 0·16
(c) 0·03
(d) 0·28
(e) 0·25
(f) 0·81

Exercise 4.6
(a) 60
(b) 60
(c) 91
(d) 15
(e) 30
(f) 60
(g) 9
(h) 21
(i) 63
(j) 31

Exercise 5.1
(a) $n = -13$
(b) $n = -20$
(c) $n = -16$
(d) $n = -7$
(e) $n = -11$
(f) $\frac{1}{16\,384}$
(g) $\frac{1}{524\,288}$

Answers 95

Exercise 5.1 (contd)
- (h) $\frac{1}{4096}$
- (i) $\frac{1}{256}$
- (j) $\frac{1}{64}$

Exercise 5.2
- (a) 162 miles
- (b) 40 000 tons
- (c) 40 miles
- (d) 60 sacks
- (e) £5120

Exercise 5.3
- (a) £4
- (b) £30
- (c) £48
- (d) £60
- (e) £25 or £26
- (f) £4·90

Exercise 5.4
- (a) 4
- (b) 3
- (c) 5
- (d) 10
- (e) 100
- (f) 64
- (g) 121
- (h) 225
- (i) 49
- (j) 144

Exercise 5.5
(a), (c), (e), (f), (i)

Exercise 6.1
- (a) 4
- (b) $3\frac{1}{2}$
- (c) $1\frac{1}{3}$
- (d) 12
- (e) $\frac{4}{9}$
- (f) 5
- (g) 11
- (h) 4

Exercise 6.1 (contd)
- (i) $6\frac{2}{3}$
- (j) $1\frac{11}{24}$

Exercise 6.2
- (a) 57
- (b) 69
- (c) 76
- (d) 33
- (e) 95
- (f) 8·3
- (g) 14
- (h) 3·4
- (i) 10
- (j) 34·5

Exercise 7.1
- (a) 46·31
- (b) 207·50
- (c) 1256·80
- (d) 193·01
- (e) 559·007

Exercise 7.2
- (a) 0·36
- (b) 110·205
- (c) 1·324
- (d) 90·362
- (e) 126·001

Exercise 7.3
- (a) 14·518
- (b) 69·8671
- (c) 39·8208
- (d) 3206·25
- (e) 19 462·0

Exercise 7.4
- (a) 26·40
- (b) 8·031
- (c) 211·0
- (d) 9·15
- (e) 80·88

Exercise 7.5
- (a) 34·5

Exercise 7.5 (contd)
- (b) 4·75
- (c) 1730
- (d) 78
- (e) 33
- (f) 975
- (g) 1690
- (h) 273
- (i) 68
- (j) £1435

Exercise 8.1
- (a) $x = 25$
- (b) 10 men
- (c) 36p
- (d) £42
- (e) 5 hours

Exercise 8.2
- (a) 9 eggs
- (b) $15\frac{3}{4}$ stones
- (c) 960 units

Exercise 8.3
- (a) £30
- (b) £153
- (c) £1123·50
- (d) £4625
- (e) £24 868·75

Exercise 8.4
- (a) 0·025
- (b) 0·04
- (c) 0·045
- (d) 0·09
- (e) 0·065

Exercise 8.5
- (a) 1·41
- (b) 0·605
- (c) 1·19
- (d) 0·00925
- (e) 0·000388
- (f) 15
- (g) 0·06
- (h) 0·025

Exercise 8.5 (contd)
(i) 20
(j) 0·002

Exercise 8.6
(a) 54%
(b) 89%
(c) 40%
(d) 80%
(e) 74%

Exercise 9.1
(a) +3
(b) +8
(c) −2

Exercise 9.1 (contd)
(d) +4
(e) −6
(f) +5
(g) +13
(h) −4
(i) +5
(j) +9

Exercise 9.2
(a) 85
(b) 13·3
(c) 368
(d) 87·5
(e) £1600

Exercise 9.3
(a) 5 hours
(b) 9 gallons
(c) 7 days
(d) 270 tiles
(e) 2 times

Exercise 9.4
(a) $3y$
(b) $-5y$
(c) $-8y$
(d) $10y$